WILDERNESS WARS

Historical Miniature Rules for Wilderness Warfare in the Eighteenth Century

by

James A. Harris & Philip S. Bock

NORMAL WARFARE PUBLICATIONS

Wilderness Wars

Historical Miniature Rules
for
Wilderness Warfare in the Eighteenth Century

by

James A. Harris & Philip S. Bock

Combatants of the American War of Independence
by
Forrest Harris

PHOTOGRAPHY
Greg Horne

PHOTOGRAPHY, GRAPHICS & COVER DESIGN
David A. Bock

ISBN 0-9748690-0-7

Normal Warfare Publications
P. O. Box 35
Normal, Illinois 61761

www.normalwarfare.com

Second Printing

This book is dedicated to our wives, Linda & Carla,
without whose love, support, and limited patience
this work would not have been possible.

Acknowledgments

PLAYTESTERS

The Acme Gamers: Rusty Graham, Forrest Harris, Tom Jaggard, Rich Nelson, Dave Sering, Roy Spencer, Jeff Sturch.

The authors would like to give special thanks to those individuals who have given us their input, encouragement and support, especially our friends at The Last Square in Madison, Wisconsin and Forrest Harris.

We would also like to thank all those individuals who have participated in the games we have presented at conventions in the various stages of the development of this rules set.

Table of Contents

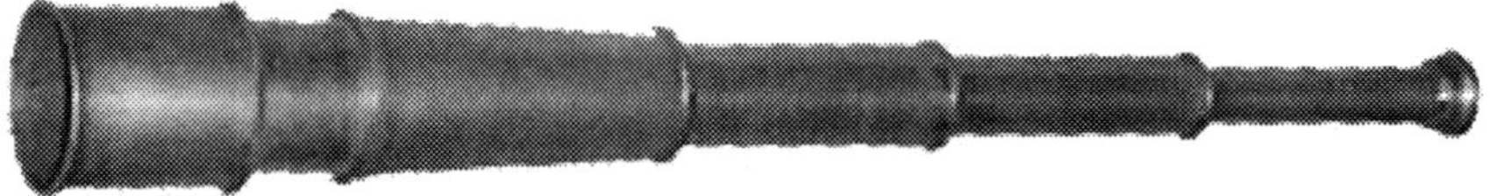

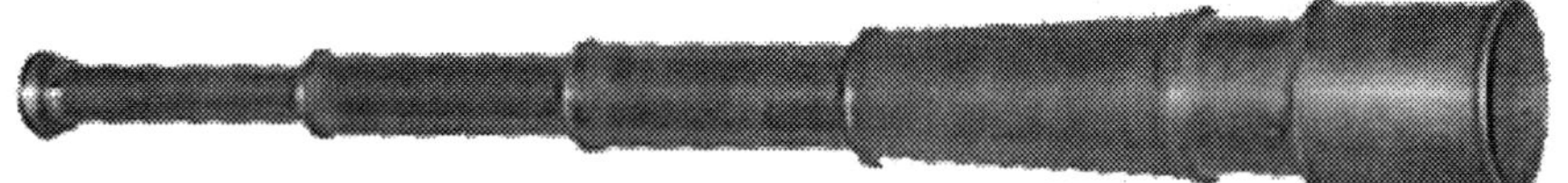

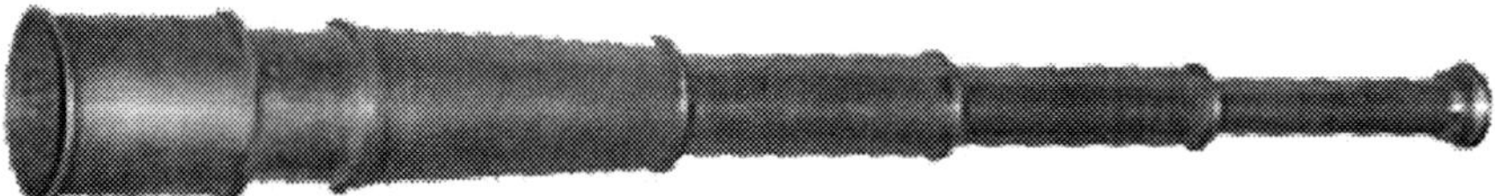

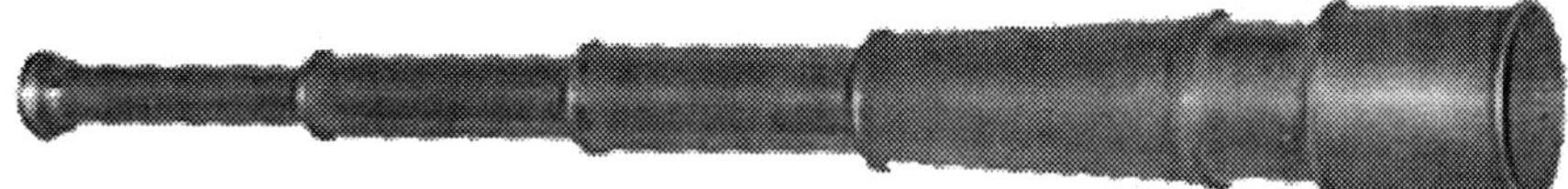

Introduction

Wilderness Wars is meant to simulate company and battalion-sized conflicts in the eighteenth century. Through the course of gaming the period for several years, it has become apparent that no current set of rules on the market accurately simulates the medium-sized conflicts of the period. Most rule sets designed for colonial warfare in this period concentrate on skirmish-level actions representing no more than several hundred men in each force.

Meanwhile, rules designed for eighteenth-century European warfare are meant to simulate battles on a much larger scale than was typically found outside of Europe at that time. *Wilderness Wars* strives to provide rules specifically designed for scenarios in which the opposing armies typically numbered anywhere from a few hundred to several thousand men each, while maintaining the unique flavor of Wars in the Wilderness.

Set Up

Important Notice to the Gamer

Although this rule system was created with 15mm figures in mind, we have discovered that it works equally as well with 25mm figures. Modification to ranges, movement rates, or any other aspect of the rules is unnecessary.

With only slight modification these rules can adapted to most mounting systems.

Setting up the Game

Scale

Each casting in *Wilderness Wars* represents ten men. Each artillery figure typically represents one or two guns. Each inch on the battlefield represents approximately 15 yards. Each turn in the game represents approximately 15 minutes of time.

Mounting Figures

The standard figure base in *Wilderness Wars* is 1½ inches wide by 1 inch deep. Most regular units are mounted five to a stand with each stand representing a company of 50 men. The size of a company may actually vary from 40 men to as many as 100 men depending on nationality or troop type. In cases of larger companies of 80 to 100 men, each company is represented by two stands with five castings each. A battalion usually contained from eight to as many as twelve companies.

Canadian militia and *Troupes de la marine* usually fought in skirmish formation. These units are usually represented by stands of four figures each. Native Americans, Rangers, and other irregulars always fought in skirmish order and are represented by stands containing two or three figures each.

All cavalry units are considered skirmish. They are represented by two figures per stand.

Artillery units are typically mounted on stands 1½ inches wide by 1½ inches deep.

The individual types of units will be described in further detail in the section entitled *The Combatants.*

Troop Class

Wilderness Wars has five basic classes of troops as listed below. These classes affect fire, morale, melee, and other factors.

A Class troops represent guard, elite grenadier, and elite riflemen or sharpshooters. B Class troops represent grenadiers, veteran regulars, and riflemen.

C Class troops represent regulars and veteran regular skirmish. D Class units consist of green units or veteran militia.

E Class units consist of irregulars with no fire discipline, militia, settlers, and mobs.

Initiative

Turn Sequence

Every turn in *Wilderness Wars* is divided into eleven phases or sequences as listed below. All players must complete each phase before moving onto the next phase.

1. Initiative
2. Charges/Countercharges
3. Normal Movement
4. Command & Control
5. Fire
6. Morale
7. Melee
8. Melee Results
9. Artillery Reload
10. Rally & Recovery
11. Army Exhaustion.

Initiative

Initiative is the process by which order of movement is determined. The player who wins the initiative by rolling the highest modified die roll has the choice of moving first or second. Initiative is rolled for at the beginning of each turn. The Commander of each opposing Army is assigned an initiative rating as illustrated in Table A.

Table A. Initiative Rating Modifiers

Initiative Rating	Modifier
A (Bold)	+2
B (Decisive)	+1
C (Competent)	0
D (Indecisive)	- 1
E (Cowardly)	- 2

The person running the game has the option of assigning the rating modifiers according to historical factors or may opt to determine the modifiers through a random dice roll as illustrated in Table B.

Table B. Random Initiative Rating

Die Roll	Initiative Rating
2	Rating of E
3-5	Rating of D
6-8	Rating of C
9-11	Rating of B
12	Rating of A

Typically, the overall commander of each army would make the initiative rolls. However, there are many instances where one may decide to have more than one commander from each side roll for initiative.

For instance, in larger battles, each corps or wing commander could be given a separate initiative rating. In battles in which allied armies under independent command are fighting together, each nation's overall commander could be required to make an initiative roll.

Another example would be a situation in which an advance column is separated from the main army and thus out of communication with the overall commander. This would force the isolated commander to act independently.

Utilization of more than one initiative roll per side could lengthen the time taken to complete the movement phase and result in a slow-down of overall game play.

Charges/Countercharges

Charge/Countercharge Phase

The only way for a unit to melee an enemy unit is through a successful charge/countercharge. The Charge/Countercharge Phase is divided into Nine Steps as listed below.

1. Determine Eligibility
2. Charge/Countercharge Declarations
   a. First Player Charges
   b. Second Player Charges/Countercharges
   c. First Player Countercharges
3. Offensive Morale Check
4. First ½ of Charge Movement
5. Defensive Morale Check
6. Defensive Reactions
7. Defensive Fire
8. Offensive Morale Check (if applicable)
9. Final Movement to Contact

Any unit that charges/countercharges will receive a +1 bonus in melee.

1. Determine Eligibility

There are certain restrictions on units wishing to charge/countercharge.

Before declaring a charge/countercharge, players must determine whether that unit is eligible.

Infantry and cavalry units are the only units eligible to charge/countercharge. Artillery may never charge or countercharge under any circumstances.

Formed infantry can never charge or countercharge mounted cavalry but may charge/countercharge any other infantry unit or any artillery unit.

Skirmish infantry may never charge/countercharge mounted cavalry but may charge other skirmish infantry.

Skirmish cavalry can charge skirmish infantry or other skirmish cavalry.

Neither skirmish infantry nor skirmish cavalry may ever charge the **front** of a formed unit or unlimbered artillery unless that unit is broken or routed.

They may declare a charge on the flank or rear of a formed infantry unit or unlimbered artillery at any time during the **charge** phase.

Indians Charging

Indians rarely charged unless they had a distinct advantage. They may never charge unless they have a 2 to 1 advantage.

2. Charge/Countercharge Declarations

Once Initiative is determined, the player moving first declares all charges. Then the player moving second declares all charges. In addition, any of the second player's eligible units being charged may now counter charge that unit if they wish.

Finally, the first player then has the option to declare any countercharges.

3. Offensive Morale Check

When a unit declares a charge/countercharge, the offensive player must first pass a morale check. Non-shock infantry units attempting to charge/countercharge have a –4 modifier to their morale roll.

Certain units that specialized in melee combat, such as Grenadiers, Highlanders and Cavalry are considered shock troops.

Therefore, these units do not suffer a –4 modifier when checking morale in this phase.

Note: If any unit declaring a charge or countercharge fails to make its **initial** morale check, that unit immediately goes into disorder.

It cannot not move **voluntarily**, make any formation changes, or changes of direction **during** the Charge/Countercharge Phase.

The exception to this rule is when the checking unit is forced to roll on the Fate Table by rolling a natural "2". (See the Morale Section for more details)

During the ensuing Movement Phase, the player that failed the charge/countercharge morale check will have the option of moving up to half its movement or using that half movement to remove the disorder.

Should a disordered unit become involved in a melee as a result of a charge/countercharge, the disorder cannot be taken off during the ensuing Movement Phase.

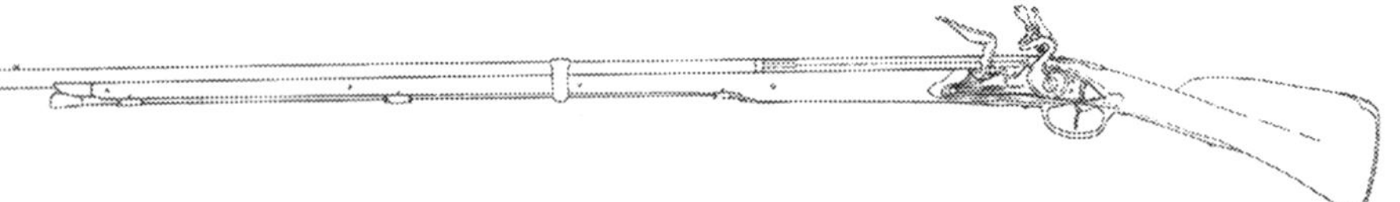

British Infantry Charging Continental Line

4. First Half of Charge Movement

On a successful morale check, the charging unit may then complete the first ½ of its charge movement.

All formation changes and changes of direction must be completed in the first ½ of this movement.

Then, any unit making a countercharge must complete all formation or facing changes within the first ½ of its movement.

At no time may a unit charge obliquely. Any direction changes must be made in the form of a wheel.

Note: If a charging unit makes contact with the enemy before spending ½ of its charge move, it must stop 1" from the defender.

5. Defensive Morale Check

After the charging unit has completed Step 4, the unit being charged must now pass a morale check.

If the defending unit passes its morale check, it may fire at close range and react as normal.

If the morale check results in a disorder, the checking unit must fire on the charge with a disorder -2 penalty.

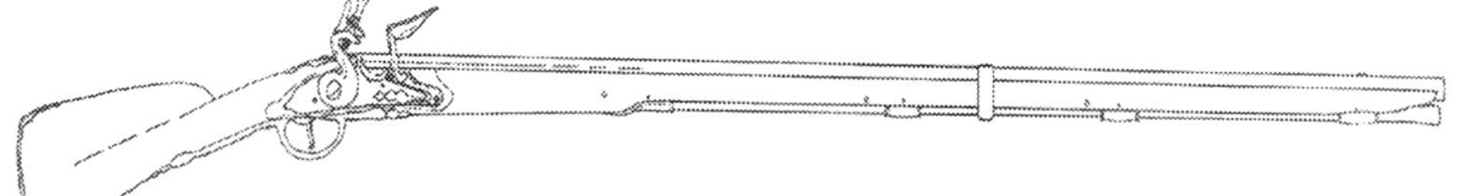

If a unit goes broken on its morale check it cannot fire on a charge. The broken unit ***must*** fall back its entire movement.

If a charge is declared against a unit that is already broken, and it passes its morale, it may stand and fire with the appropriate modifiers.

If the morale check of the defending unit results in a rout, that unit must immediately make a full rout move in a direct line away from the charging unit. It also suffers one casualty.

Routed units that are charged automatically rout again, taking a full rout move away from the charge prior to contact and suffer one casualty.

6. Defensive Reactions

Defending units that were not struck in Step 4, that successfully passed their morale check, may now choose to make defensive reactions that do not involve fire.

Any disordered, broken, or routed unit cannot perform any reactions to the charge/ countercharge.

Infantry

A formed infantry unit that is charged on the flank or rear may attempt to use up to ¼ of its movement to change formation, face the charger, refuse the flank.

Only infantry in line formation may refuse the flank. A unit would do so by pivoting the end stand in the line to face the charge.

Cavalry

A cavalry unit that is charged on the flank or rear may use up to ¼ of its movement to change formation, face the charger, or attempt to evade.

Evade: A cavalry unit may attempt to evade by moving away from the charging unit in any direction using up to ½ of its movement.

If a unit attempts to evade, it is subject to pursuit as the charging cavalry unit has targeted them.

Pursuit is the **only** exception to the rule that all direction changes must be made in the first ½ of charge movement.

Note: Only cavalry units can pursue.

For details on Pursuit see Step 9, Final Movement to Contact.

Skirmish Infantry

Skirmish Infantry being charged by formed infantry must evade up to their full movement, subject to terrain modifiers. They may fire with appropriate penalties for movement.

Skirmish infantry being charged by cavalry in the open must evade up to ½ of their movement, subject to terrain modifiers. They may fire with appropriate penalties for movement.

They have the option of stopping behind a formed unit they have interpenetrated or once they reach cover.

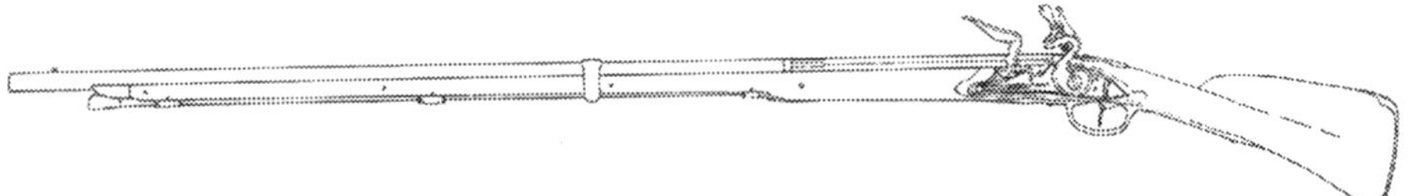

Artillery

Limbered artillery may attempt to evade up to ½ of its movement away from the charging enemy.

Note: Any unit that evades may not voluntarily move for the remainder of the turn.

7. Defensive Fire

Defending units that successfully passed their morale check now have the option to fire.

Note: If the charging unit makes contact in the first half of charge movement, the **only** reaction available is to stand & fire. You cannot run if you've been hit.

Infantry

If the defending unit passes its morale check, it may fire when the charging unit reaches close range.

Any infantry unit that became disordered as a result of its morale check may now fire with a -2 penalty.

A previously broken unit that successfully passes its morale may now fire with a -3 penalty.

6 Pound Gun

Artillery

Limbered Foot artillery that is being charged does not have the option of unlimbering and firing.

Light Horse artillery may unlimber and fire taking the penalties for both movement and opportunity fire. They must then stand with their guns and fight in melee.

Unlimbered artillery that is being charged in the front has the option of firing and joining in melee or firing and fleeing (abandoning their guns and moving up to ½ of their movement).

If they choose to flee, they must pay a -4 modifier for opportunity fire with movement.

Unlimbered artillery being charged on the flank or rear has the option to either pivot and fire or abandon their guns.

If they wish to pivot and fire, they are subject to the firing penalties for both movement and opportunity fire with movement. They cannot abandon their guns and must stand and fight in melee.

8. Offensive Morale Check

If a charging unit takes any casualties as a result of fire, including opportunity fire, the charging unit must take another morale check. If successful, it may then complete the charge and move to contact.

The charging unit will only take one morale check for casualties on a charge, regardless of the number of units that fire upon it.

If the charging unit fails this Offensive Morale Check, follow the normal morale procedures.

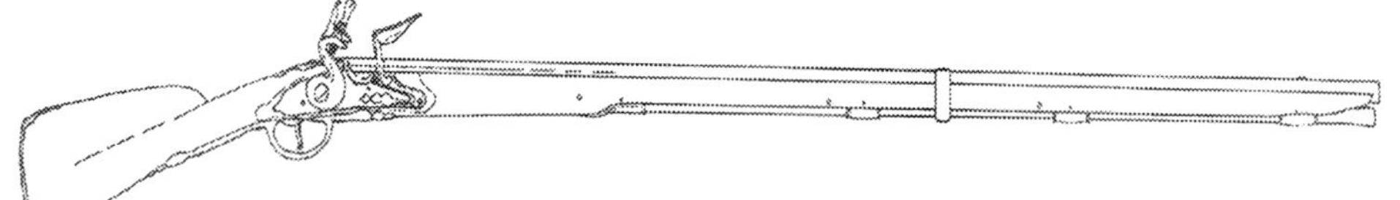

9. Final Movement to Contact

Once the charging/countercharging unit has successfully completed all morale checks, it may now attempt to move to contact.

If charging/countercharging units did not make contact during the first half of their movement, each unit moves **sequentially**, beginning with the charger, using ¼ of their movement at a time until contact is made or they both reach their maximum charge move.

It is possible for a charging/countercharging unit to use its full charge movement and not come into contact with the enemy.

If a charging/countercharging cavalry unit does not come into contact with the enemy, its horses are considered to be "blown" and it immediately goes into disorder.

If the defending unit withdraws for any reason, the charging unit then has the option of taking the ground vacated by that unit, continuing up to the full extent of the remaining charge movement or pursuing.

Pursuit: A charging cavalry unit has the option to target an evading unit. It may pursue that unit up to its full charge movement.

At no time may a unit use an oblique move while pursuing—the charging unit must wheel in order to change direction.

Conclusion

Once the charge movement is completed and contact is made, these units become locked in melee. They may not perform any other actions until the Melee Phase.

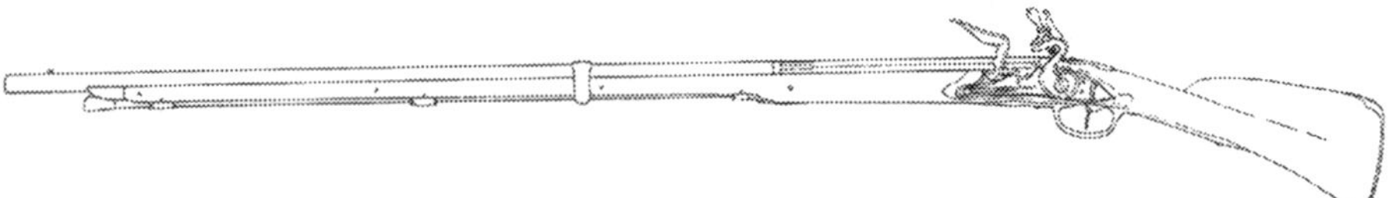

Movement

Introduction to Movement Phase

The Movement Phase begins after all charges have been resolved. It is divided into two steps.

1. First Player Movement
2. Second Player Movement

Once the first player has completed **all** movement, the second player then moves.

This chapter is divided into the following sections:

- Base Movement Rate
- Facing and Formation Changes
    - Expanding & Collapsing Frontages
    - Skirmish Formations
    - Artillery Formation Changes
    - Pass Through
- Terrain Effects
- Opportunity Fire
- Disordered, Broken & Routed Units

Base Movement Rate

The player moving first, moves his units at the base movement rate listed in Tables C and D. This represents the maximum distance a unit may move in a turn, before applying movement modifiers.

Normal movement refers to the distance a unit may move per turn, after modifiers have been applied.

Table C. Base Movement Rate Table

Formation	Infantry	Cavalry
Line	12"	24"
Column	16"	30"
Line Charge	16"	36"
Column Charge	20"	42"
Skirmish	24"	36"
Road Column	24"	54"
Rout	12" + 2 die	24" + 4 die
Skirmish Rout	12" + 4 die	24" + 4 die

Table D. Base Artillery Movement Rate

Size	Field	Road	Prolong
3 pounds or less (light)	20"	28"	5"
4 to 6 (light)	20"	28"	4"
7 to 10 pounds (medium)	18"	26"	3"
11 to 15 pounds (heavy)	16"	24"	2"
17 pounds or greater (siege)	8"	14"	Pivot Only
Howitzers and mortars			
3-inch or less howitzers	20"	28"	5"
4- to 6-inch howitzers	20"	28"	4"
7- to 10-Inch howitzers	18"	24"	3"
11- to 15-Inch howitzers	16"	24"	2"
17-inch or greater	8"	14"	Pivot Only

Facing and Formation Changes

An infantry unit may be in line, column, or skirmish formation. A cavalry unit may be in line, double line, column, or skirmish formation. A unit that changes formation is subject to a movement penalty dependent upon its morale class. The unit's command stand is the base from which all formation changes are made.

A unit in line may make an about-face without paying a movement penalty. If a unit makes a second about-face in the same movement phase, it must pay the penalty for a formation change. Columns are not allowed to about-face but may "snake".

An infantry unit may retrograde (back up while maintaining their original facing) at ½ its normal movement. Cavalry may not retrograde.

A unit may move in a wheeling fashion. This is done by having one end of the unit remain stationary while the other flank moves forward or backward.

A unit in line may also make an "oblique" move by maintaining its facing while moving forward at an angle of up to 45 degrees. Penalties for moving obliquely are determined by morale class as illustrated in Table E.

Note: Mounted officers & messengers move as skirmish cavalry, while officers & messengers on foot move as irregular skirmish infantry.

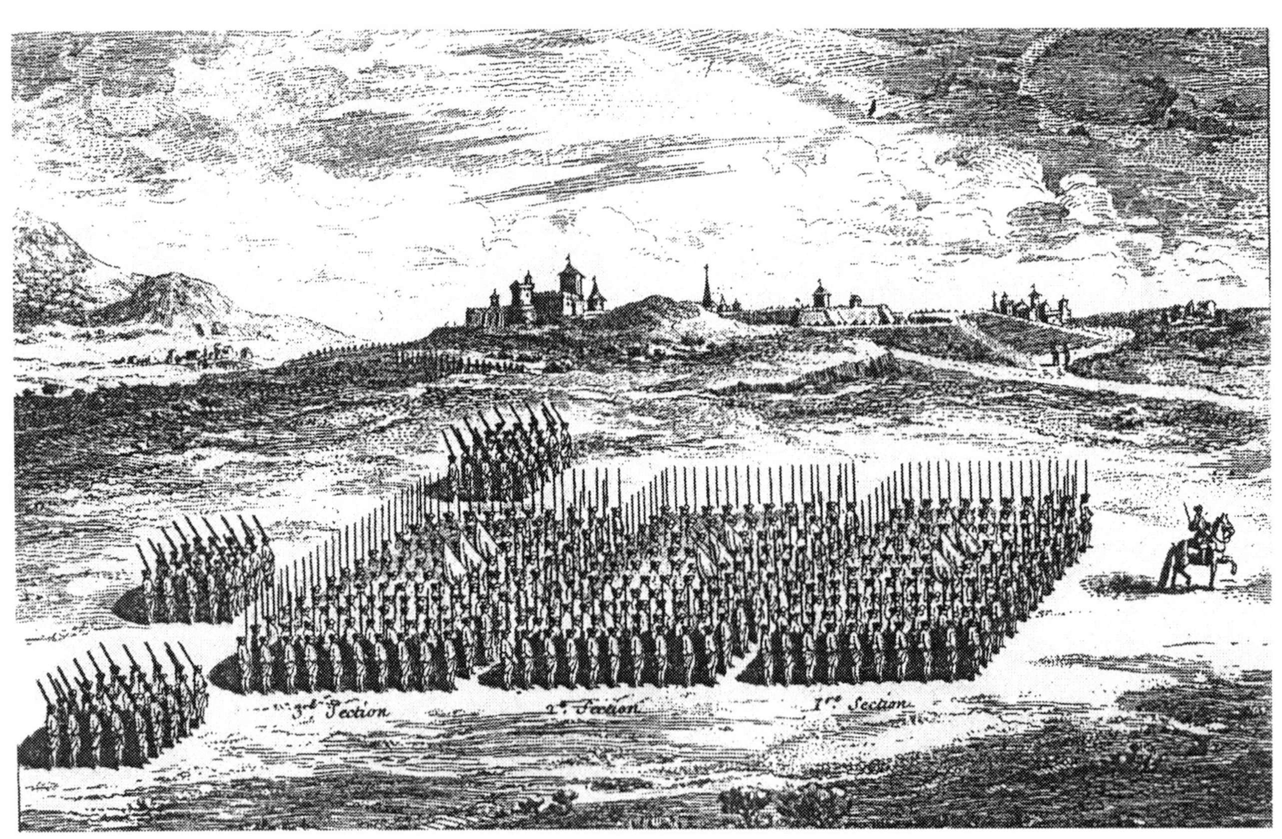

Table E. Troop Quality Movement Penalties

Troop Class	A	B	C	D	E
Retrograde	½	½	½	½	½
Oblique	NE	NE	½	½*	½*
Formation Change	¼	¼	½	Full	Full

* Unit becomes disordered NE = No effect

Expanding & Collapsing Frontages

Only Cavalry regiments can expand and collapse frontage. Cavalry Regiments can form into two types of line formations—Single line or double line. Double line allows a larger unit to form into line in an area of relatively small frontage due to terrain restrictions.

As a cavalry regiment moves forward it is allowed to expand or contract its frontage at the rate of one stand for every 4" of forward movement.

When expanding or collapsing, a cavalry unit must line up to either the left of the command stand (Natural Order) or to the right of the command stand (Inverse Order). The command stand must always be the lead unit in a column.

Skirmish Formations

There are two types of skirmish units—regular and irregular.

Regular skirmish units are those that move in an organized skirmish line. Most cavalry units, regular skirmish infantry, some organized militia, provincial regiments, etc. would fall into this category.

Regular skirmish units have the capability of marching in column or line. Some, such as militia may also form into regular lines but rarely did so.

They move and act like regular formed units but do not become disordered when moving through wooded or rough terrain.

Regular skirmish units pay ¼ movement to make a formation change without becoming disordered regardless of morale class.

Regular skirmish units that wish to move laterally may make a front-to-flank maneuver.

As an example, they may expend ¼ of their move to make a 90 degree facing change going into single file. They would then move forward as a column for ½ their movement and then make another 90 degree facing change expending the final ¼ of their movement.

Irregular skirmish units are those that typically move in a mob fashion with little or no formal training. Examples of this type of unit would be Native American warriors, ranger units, groups of settlers, or street rabble.

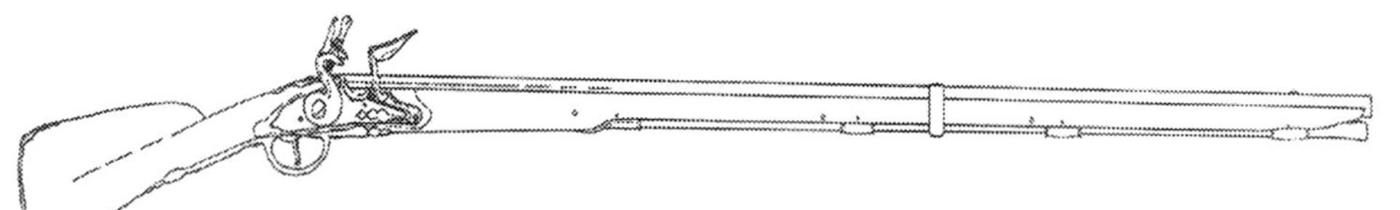

Irregular skirmish units can never march in column formation but may march in file.

Irregular skirmishers do not have conventional formations so they may move in any direction at will without paying a penalty for formation change.

Artillery Formation Changes

Artillery may perform three functions: limber/ unlimber, movement, and fire. All foot artillery may perform two of these actions in any turn.

Artillery may limber, unlimber, pivot or make a facing change without suffering a movement penalty. Artillerists may also prolong by manually moving the gun.

Pass Through

Friendly units may pass through other friendly units without causing disorder providing one of the units remains stationary.

Skirmish units are the exception to this rule. They may move without penalty even when passed through by friendly units.

If two formed units pass through each other and both moved they will each go into disorder.

A disorder that is obtained through normal movement will automatically be removed during the rally phase.

Note: This rule does not apply to units that are forced back or are in rout.

For further details, see the section on Morale.

Terrain Effects

Units are affected by the type of terrain through which they move. The base movement for a particular movement represents the maximum distance that they can move through an open field or meadow in a given turn.

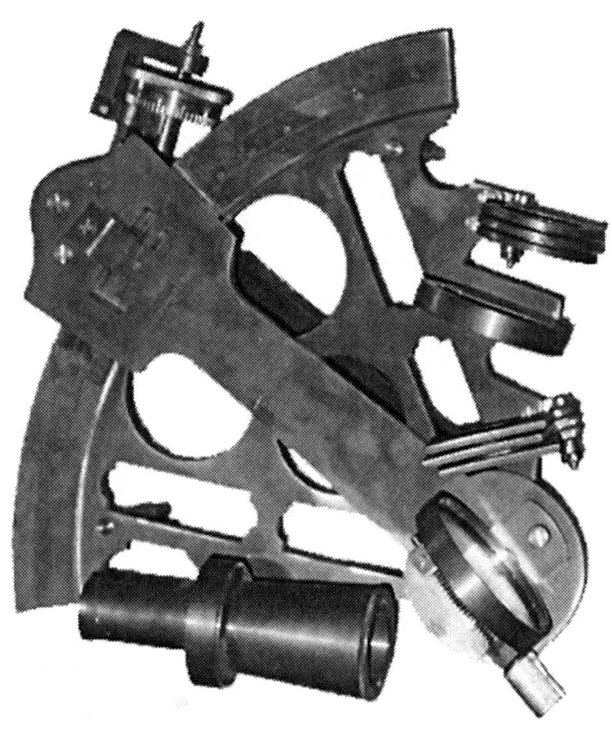

Brass Sextant

Woods

There are two types of woods in *Wilderness Wars.* Light and heavy woods affect each type of unit with different movement penalties.

Regular infantry units move ½ at their movement rate for light woods and ¼ of their movement for heavy woods.

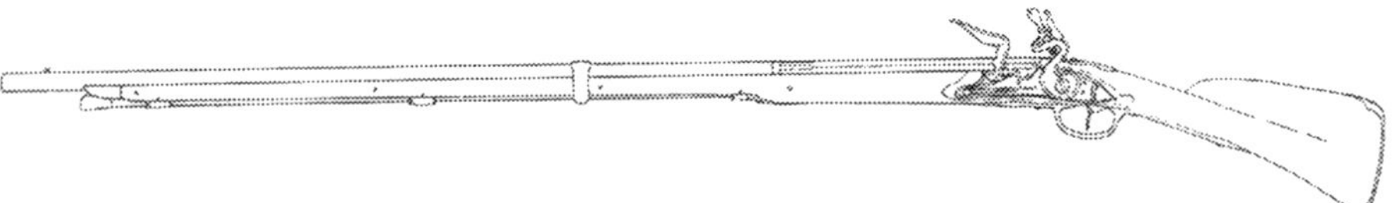

Regular infantry units moving in wooded terrain immediately go into disorder. This disorder is considered a movement penalty and automatically comes off during the Rally & Recovery Phase.

Regular skirmish infantry moves through light woods at ½ speed and heavy woods at ¼ speed.

Note: The benefit to having regular skirmish units such as provincial and militia regiments is that they do not suffer disorder when moving through woods.

Irregular skirmish infantry pays no movement penalty in light woods. These units move through heavy woods at ½ speed. They do not go into disorder due to movement through wooded terrain.

Cavalry units move through light woods at ¾ speed. Cavalry cannot move through heavy woods. Since all cavalry units in *Wilderness Wars* are skirmish, they do not suffer disorder when moving through light woods.

Artillery can move through light woods at ¾ speed. These units do not suffer disorder when moving through light woods.

Artillery cannot enter heavy woods but may set up at the edge of the woods and receive a bonus for cover when attacked or fire upon.

However, artillery does go into disorder when setting up on the edge of heavy woods. This disorder is considered a movement penalty and automatically comes off during the rally and recovery phase.

Surveyor's Transit and Tripod

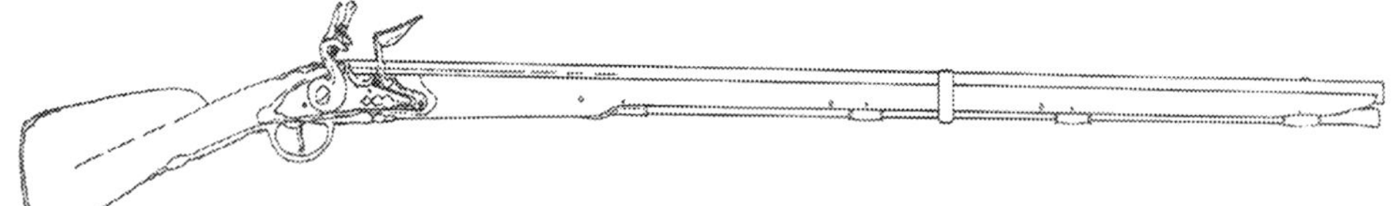

Hills

Normal low rolling hills or rises do not affect movement. However, there is a penalty for climbing steep hills. Only artillery is penalized when moving down a steep hill. Hills do not disorder units. See Table F for details.

Water

Passage through water also effects movement depending on the depth of the water. Regular infantry and artillery are disordered when moving through all types of water.

All units are disordered when moving through fordable water above the knee. Deep water is obviously impassable other than by bridge or boat.

Units standing in water above their knees cannot fire.

Linear Obstacles

Linear obstacles such as fences, small walls, or hasty works have an affect on infantry and cavalry movement as seen in Table F. It takes artillery an entire turn to move across a linear obstacle.

Only regular infantry units go into disorder when crossing a linear obstacle.

Rough Ground

Rough ground is categorized as any area that contains debris, rubble, or is a freshly plowed field.

Towns & Villages

Only skirmish units can go through a town in line. Regular infantry, cavalry, and artillery must pass through a town in column. Regular infantry or artillery may form line on the edge of a town using the buildings as a defense.

Note: At a scale of 15 yards to the inch, most buildings on the tabletop are more likely to represent one or two buildings as opposed to an entire town or village.

The above rule can be modified at the player's discretion. This will accommodate those Gamers who wish to have the ability to charge into a building.

Line of Sight is 6" in light woods and 3" in heavy woods. The line of sight across open fields is unlimited.

See Table F for terrain effects.

Seneca Long House

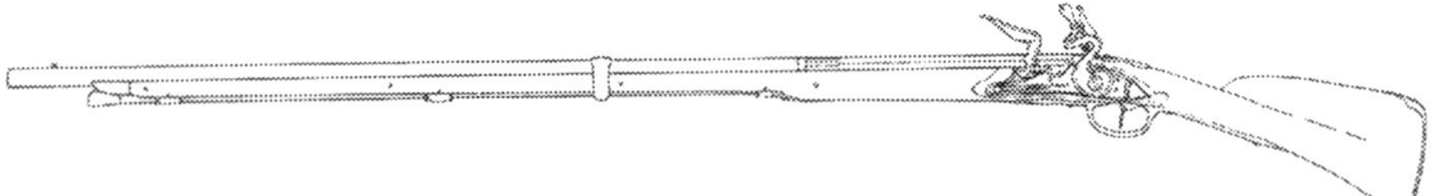

Table F. Terrain Effects

Terrain type	Irregular Skirmish	Regular Skirmish	Regular Infantry	Cavalry	Artillery
Light Woods	NE	½	½	¾	¾
Heavy Woods	½	¼	¼	N/A	N/A
Low Rolling Hills	NE	NE	NE	NE	NE
Steep Hills	¾	½	½	½	½ up ¼ down
Shallow Water (below the knee)	¾	¾	¾	¾	¾
Fordable Water (above the knee)	½	½	½	½	½
Deep Water		Impassable			
Marshes	½	½	½	N/A	N/A
Bridges	NE	½	½	½	NE
Towns	NE	½	½	½	NE
Linear Obstacle	¾	¾	¾	½	1 turn
Rough Ground	¾	¾	¾	½	½
Disordered Unit	½	½	½	½	½
Broken Unit	¼	¼	¼	¼	¼

NE = No Effect
N/A = Not Applicable. These units cannot move through this terrain.

French Troops & Their Indian Allies Move Through the Wilderness

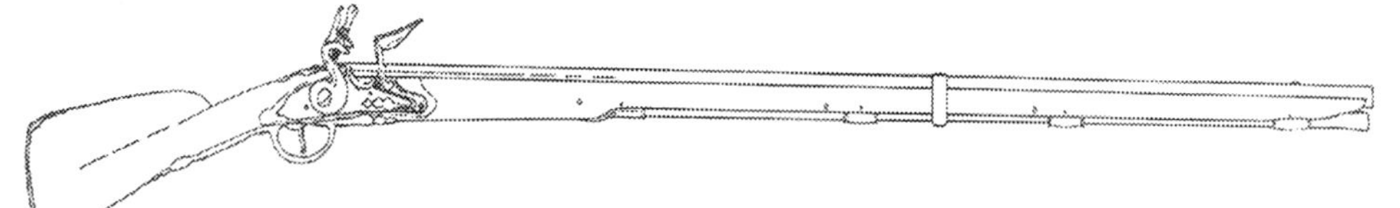

Opportunity Fire

Opportunity fire is one of the most abused rules in wargaming. These rules may at first glance appear complicated but the game designers feel they have devised a system incorporating opportunity fire that does not penalize those units that are required to move first during the movement phase.

If during the course of the movement phase, a unit moves within or through an opposing unit's field of fire the opposing player may choose to take opportunity fire. The player choosing to take opportunity fire must declare so during the opposing player's movement.

Once opportunity fire is declared, the moving unit must halt until the opportunity fire is completed. Casualty Dice are calculated in the usual manner but the firing unit is subject to an additional –2 kill modifier for opportunity fire.

If the firing unit has already moved or declares an intention to move later during the turn, it must fire with a –4 penalty.

See the fire phase section for details on calculating fire modifiers.

If the moving unit receives casualties as the result of opportunity fire, it must immediately take a morale check. If it passes, it may proceed with the remainder of its movement as normal. If, however, it fails it must immediately follow the morale results as rolled.

See the Morale Results section for details.

Disordered, Broken & Routed Units

Disordered units are limited to ½ their normal movement minus any applicable terrain modifiers.

Disordered units may choose to rally during the movement phase at a cost of ½ their normal movement. Therefore, a disordered unit choosing to rally may not move voluntarily in that turn.

Note: Removing the disorder from a unit is considered movement for purposes of fire.

Broken units move at ¼ of their normal movement rate.

Routed units may not voluntarily move.

Note: Units that have received **75%** casualties cannot advance toward the enemy. However, these units may advance if they are behind friendly lines and may provide rear support for units in front of them.

Command & Control

Command and control is the phase in which it is determined whether a unit is within effective command radius of its commanding officer. It is not the intention of *Wilderness Wars* for command and control to be the decisive factor in a game, but needs to be taken into consideration.

Each army has an overall commander with a command radius of 18". In addition, any sub-commanders have a 12" radius.

Any unit that is more than 12" away from its commanding officer is considered out of command and control.

A commanding officer must be within line of sight for a unit to be within command and control. Friendly units do not block line of sight for this purpose.

Note: A unit that is out of command and control during this phase will be subject to the –2 penalty for all subsequent morale checks until the next command and control phase.

For example, a unit that begins a charge **in** command and control but then finds itself more than 12" from its commander during the melee phase, will be subject to a –2 morale penalty.

Likewise, a unit may attempt to charge while out of command and control. If successful and if after movement, the unit's commander is now within 12", the unit would be back in command and control for the melee phase.

A commanding officer that is attached to a unit may not exert command and control over any other unit.

Fire

Introduction to Fire Phase

The two types of fire in *Wilderness Wars* are small arms fire and artillery fire. All fire is simultaneous.

This chapter is divided into three sections.

- Small Arms Fire
    - Casualty Dice Determination
    - Small Arms Modifiers
- Artillery Fire
    - Artillery Modifiers
    - Artillery vs Fortifications
- Targeting Officers & Messengers

Each unit is assigned a fire class based on its quality. A unit's fire class represents its relative skill, fire discipline, tactics, and weaponry as illustrated in Table G below.

There is no pre-measurement for small arms fire in *Wilderness Wars*.

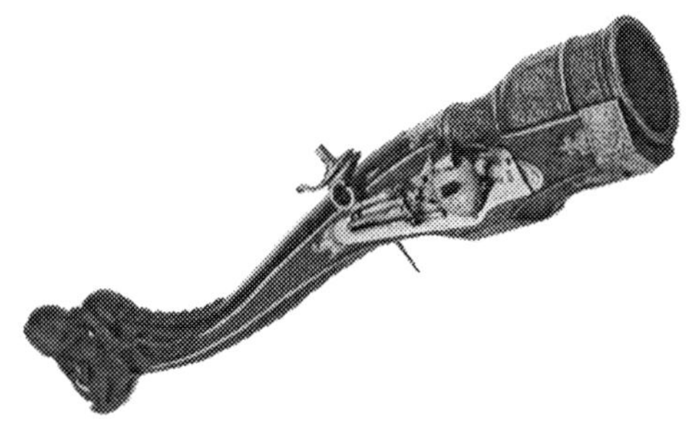

French Grenade Launcher circa 1759

All players must declare targets at the beginning of the fire phase for all units that will be firing. Units that took opportunity fire during the movement phase are prohibited from firing again in the fire phase.

Most units may fire up to a 45° angle from the front of the stand. The only exception are Irregular skirmishers, who may fire from all sides.

Note: Units standing in water above their knees cannot fire.

Table G. Troop-Type Modifiers

Class	Modifier	Unit Type
A	+2	Guard Class, elite riflemen, sharpshooters.
B	+1	Veteran British, grenadiers, riflemen.
C	+0	Regular troops, veteran regular skirmish.
D	-1	Green regulars, veteran militia.
E	-2	Irregulars with no fire discipline, militia, settlers, mobs.

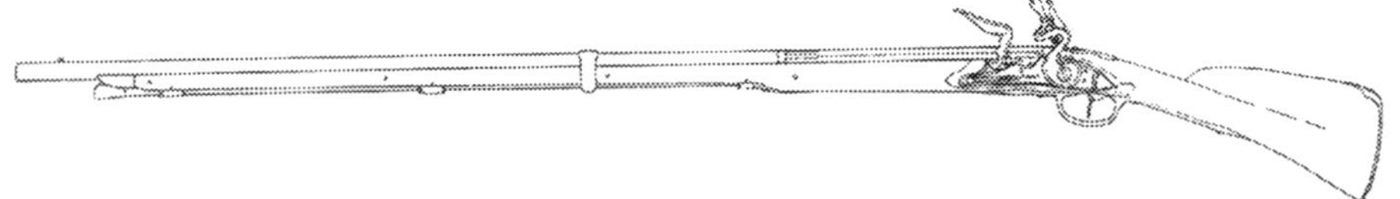

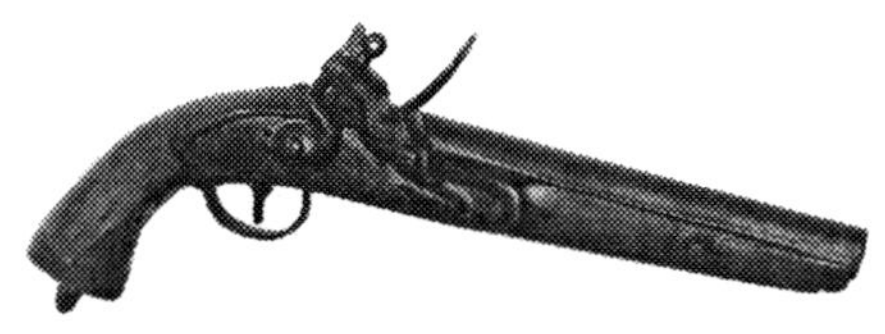

Small Arms Fire

Ranges for small arms fire are listed on Table H.

Table H. Small Arms Fire Ranges

Weapon	Short	Normal	Long
Musket	3"	6"	12"
Rifle	5"	10"	18"

Casualty Dice Determination

When firing small arms, the player must first determine the number of casualty dice. The amount of casualty dice is determined by the firing unit's strength factors.

Each stand with four or more figures equals one strength factor. Stands with three figures or less, i.e. Skirmishers, are equal to ½ of a strength factor. Fractions are rounded up. All units roll a minimum of 2d6 and a maximum of 4d6.

Thus, a unit's strength factors decreases as it loses stands. All casualties must be taken off of one stand until that stand is eliminated.

Once the number of casualty dice are determined, the player rolls the dice, factors in the small arms fire modifiers as listed in Table I, then divides by "5" to determine the number of casualties actually inflicted.

The firing player causes one casualty for each resulting multiple of "5" discarding the remainder. For example, if the modified die roll results in a "12", the firing player would inflict two casualties on the opposing unit.

Table I. Small Arms Fire Modifiers

Modifier Description	Modifier
Short Range	+2
British Regular first fire	+2
Vs. Column or Mob	+2
A class Troops	+2
Target unit is in the open	+1
Vs. Formed troops in line	+1
Each Strength Factor over 4	+1
Vs. Flank	+1
Non-British Regular first fire	+1
Vs. Cavalry	+1
B class Troops	+1
Normal Range	0
Vs. Light woods & wooden fences	0
C class Troops	0
Formed unit moved under ½	-1
Vs. Heavy woods	-1
Highlanders firing	-1
D class Troops	-1
Disordered	-2
Only 1 Strength Factor	-2
Skirmish unit moved more than ½	-2
Vs. trenches & stone fences	-2
Vs. wooden palisades & buildings	-2
Opportunity fire without movement	-2
E class Troops	-2
Broken	-3
Long Range	-3
Formed unit moved more than ½	-3
Vs. stone buildings or fortifications	-3
Officer or messenger	-4
Opportunity fire with movement	-4

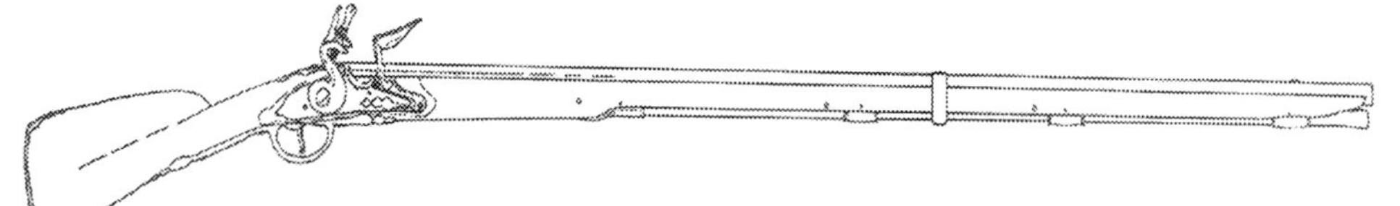

Small Arms Fire Modifier Descriptions

Below are explanations for some of the more important small arms fire modifiers.

Troop Class: Guard and A Class troops are the elite soldiers. There are normally only a few in each army. Some grenadiers fall into A class. They receive a +2 firing bonus.

B class troops represent veteran seasoned troops. Some regular soldiers and experienced woodland fighters fall into this category. They receive a +1 firing bonus.

C Class troops are regular drilled troops who have seen little or no combat. Some seasoned provincial regiments would fall into this category. As such they receive no firing bonus.

D class troops represent newly recruited inexperienced soldiers such as better quality militia or provincial regiments. They fire at a –1 penalty.

E class troops are those with little or no military training and extremely low morale. Examples of E class troops would be low-quality militia, settlers, reluctant Native American warriors, mobs, or armed working parties. They fire at a –2 penalty.

Range Modifiers: Units firing at Short Range receive a +2 firing bonus. Those firing at Long Range are subject to a –3 penalty. Those firing at Normal Range receive no modifier.

First Fire Bonus: British and French regulars were highly trained career soldiers. As such their discipline and field tactics gave them an advantage over other troops.

To reflect the increased effectiveness of their initial volleys, British regulars receive a bonus of +2 the first time they fire in a battle.

Other regulars receive a +1 bonus for their first fire.

Highlanders (French & Indian War only), fire with a –1 penalty. This reflects their relative lack of fire discipline when compared with other British regular troops. This defect is more than made up for by their superior melee capabilities.

Formation Effects: Certain formations can be fired upon with a bonus. Most battles in Eighteenth Century North America and in a few other colonies involved large numbers of skirmish troops.

Important: In order to streamline the firing system, there is no penalty for firing at skirmish or artillery units. Instead, any unit firing at a formed unit receives a +1 bonus. **Open order units thus become the norm for fire.**

A unit firing at a column or mob receives a +2 bonus. A unit firing on the flank of another unit receives a +1 bonus. These two modifiers are not cumulative because firing into the front or rear of a column will cause more casualties than firing into the flank.

A unit firing at Cavalry also receives a +1 bonus.

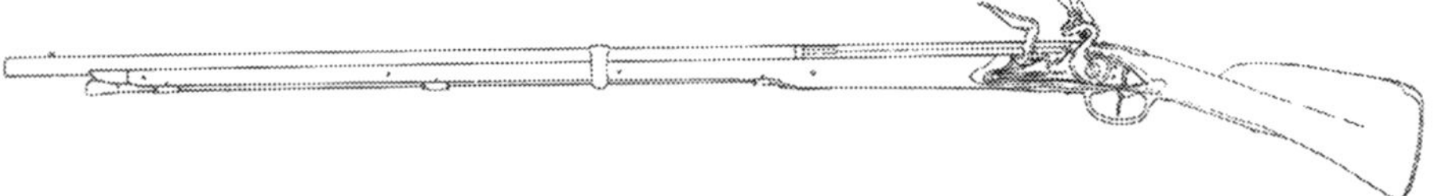

Terrain Effects: Most battles in Eighteenth Century North America took place on the frontier, in wooded areas.

Again, in order to streamline the firing system, the decision was made not to impose a penalty for firing into light cover.

Instead, a unit firing at a unit in the open would receive a +1 bonus. **Thus light cover becomes the norm for fire.**

There is a –1 penalty for firing at a unit in heavy woods.

Buildings, Fortifications & Fences: Normal and long range small arms fire is ineffective against wooden buildings, trenches, wooden palisades, stone fences, or any type of fortification. Units in these types of cover can only be fired upon with small arms at short range.

There is a –2 penalty when firing at short range into wooden buildings, trenches, wooden palisades, or stone fences. There is a –3 penalty for firing at short range at troops behind stone or earthen fortifications.

There is no penalty for firing at a unit behind a wooden fence, as it is considered light cover.

Movement Penalties: A formed unit that moved less than ½ its modified movement may fire at a –1 penalty. If a formed unit moves more than ½ its modified movement, it fires at a –3 penalty.

A skirmish unit may move up to ½ its modified movement rate without suffering a firing penalty. However, if it moves more than ½ its movement it may still fire, but at a –2 penalty.

Morale Effects: Disordered units fire at a –2 penalty. Units that are Broken may also fire but do so at a –3 penalty. Routed units cannot fire.

Strength Factors: A unit that has only 1 strength factor may fire using the minimum two dice but does so at a –2 penalty.

A unit that has five or more strength factors gains an additional +1 bonus for every strength factor over four.

Opportunity Fire: A unit may take opportunity fire at a –2 penalty. If the firing unit moved or will move during the same turn it must take a –4 penalty. See the section on opportunity fire in the Movement Phase Chapter for details.

Officers & Messengers were often targeted by native and colonial troops. A unit may fire upon Officers or Messengers under certain circumstances, but does so at a –4 penalty.

See the section entitled Targeting Officers & Messengers later in this Chapter for details.

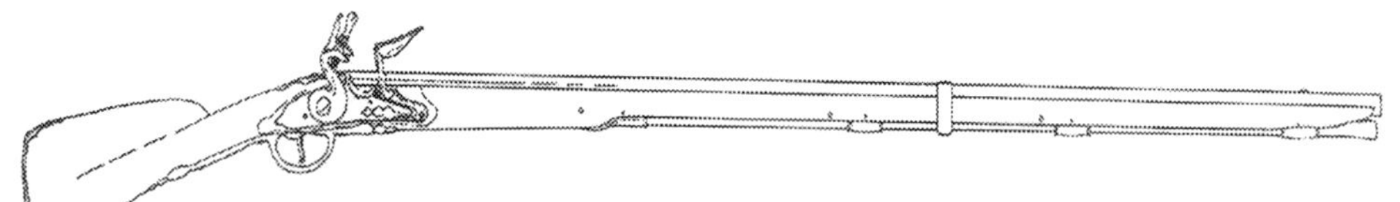

Artillery Fire

When firing artillery, first determine whether to fire ball or canister. This is done when the artillery piece is first unlimbered and during any subsequent Artillery Reload Phase.

Artillery units may fire up to a 45° angle from the front of the unit.

As with small arms fire, there is no pre-measurement for artillery fire in *Wilderness Wars.*

Roll 2d6 for each stand, modified for the size of the gun per the Artillery Size Table J.

Table J. Artillery Size Table

Size	Poundage
Swivel, Wall guns	up to 1
Light	2 to 5
Medium	6 to 10
Heavy	11 to 18
Siege or position guns	19 +

For howitzers and mortars barrel size is equivalent to poundage.

Artillery ranges can be found in Table K.

Once the artillery size and range modifiers have been determined, artillery fire is resolved in the same manner as small arms fire, using Artillery Fire Modifiers Table L.

Table K. Artillery Fire Ranges

Gun Type	Canister	Normal	Long	Extreme
Swivel, Wall Guns	3"	5"	10"	20"
2 to 4 pounders	10"	20"	40"	80"
5 to 7 pounders	11"	22"	44"	87"
8 to 15 pounders	13"	27"	54"	107"
16 pounders plus	15"	29"	58"	116"
Howitzers	8"	15"	30"	60"
Mortars	N/A	15"	30"	60"

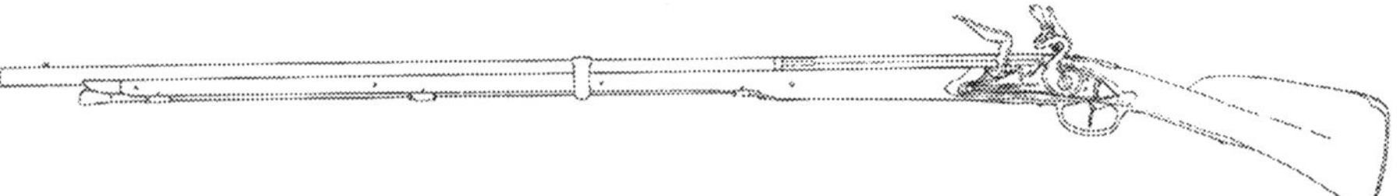

Table L. Artillery Fire Modifiers

Artillery Fire Modifiers	Modifiers
Siege Gun	+3
Short Range	+2
Canister	+2
Vs. Column or Mob	+2
A class Gunners	+2
Heavy Gun	+2
Target is in the open	+1
Vs. Formed troops in line	+1
Regular first fire	+1
Vs. Flank	+1
Vs. Wooden fence, due to splintering	+1
Vs. Cavalry	+1
B class Gunners	+1
Medium Gun	+1
Normal Range	0
Vs. Light woods	0
C class Gunners	0
Light Gun	0
Firing unit moved	-1
Vs. Heavy woods	-1
Vs. Wooden palisades & buildings	-1
D class Gunners	-1
Swivel & Wall Guns	-1
Disordered unit firing	-2
Opportunity fire without movement	-2
Vs. Trenches & stone fences	-2
Indirect fire (Howitzers & Mortars)	-2
E class Gunners	-2
Broken unit is firing	-3
Long Range	-3
Vs. Stone buildings or fortifications	-3
Opportunity fire with movement	-4
Officer or messenger (canister only)	-4
Extreme Range	-5

Artillery Fire Modifier Descriptions

Most of the modifiers for artillery are the same as those for small arms fire with a few exceptions described below:

Canister: If an artillery piece fires canister at short range it receives a bonus of +2 to reflect the increased damage it would inflict. This bonus would be in addition to other modifiers, including +2 for short range.

Canister may be fired at up to full normal range. However the +2 canister bonus only applies when fired at short range. Howitzers may not fire canister indirectly.

Wooden Fences: Artillery firing at units behind a wooden fence receive a +1 bonus to reflect the increased damage inflicted by the splintering wood.

Indirect Fire: Howitzers may choose direct or indirect fire. They may not fire indirectly at short range. Mortars always fire indirectly. Mortars, and Howitzers that fire indirectly, suffer a –2 penalty.

Indirect fire may negate some modifiers for targets in certain types of cover.

Movement Penalty: Any artillery piece that limbered, unlimbered, prolonged or pivoted during the movement phase will suffer the movement modifier penalty.

Opportunity Fire: Limbered artillery may not take opportunity fire.

An artillery piece cannot pivot to take opportunity fire. It may only take opportunity fire on units that pass within its 45° arc of fire.

For further details, see the section on Opportunity Fire in the Movement Phase.

Officers & Messengers: Artillery may only fire at an officer or messenger with canister.

Terrain Effects: Artillery cannot unlimber or fire while in any type of water.

Extreme Range: Artillery may fire at extreme range. However, it is relatively ineffective and fires at a -5 penalty.

Artillery in the Eighteenth Century was not used as effectively as it was in the Napoleonic Period. The reforms of Jean-Baptiste Gribeauval had not yet been implemented.

As such, artillery fire in *Wilderness Wars* reflects military procedures in place prior to the Gribeauval reforms and is not as effective as it was in later years.

French Artillery and Swivel Gun Defending the Redoubt

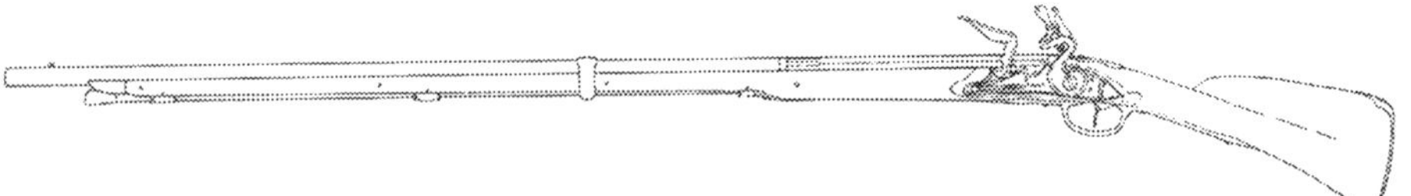

Targeting Officers & Messengers

Individuals during any period of warfare are not exempt from being killed, wounded or captured.

Most rule sets allow an individual officer to only become susceptible when they are attached to a particular unit that has either come under fire or is involved in melee. *Wilderness Wars* wishes to address that historical inaccuracy.

Attached Officers: An officer that is attached to a unit that comes under fire or is involved in melee, is vulnerable to being killed, wounded, or captured.

If any casualties are inflicted on a unit with an attached officer, the attacking player rolls 1d6. On an unmodified roll of 5 or 6, consult the Officer & Messenger Casualty Table M and roll again. The results are applied immediately.

Table M. Officer & Messenger Table

Dice Roll	Results
1	Hat shot off. No effect.
2	Slight wound. Out for one turn.
3	Serious wound. Out for two turns.
4	Horse shot, falls on rider. Out for three turns. Captured if in melee.
5	Mortal wound. Officer dies after this turn
6	Shot dead. Hacked to pieces if in melee, body unrecognizable.

Unattached Officers: In addition, if an Officer or Messenger is in a position where they are the prime targets, they may be fired upon by small arms fire. An artillery battery may select an individual as a prime target only when firing canister.

Note: Should a firing unit have any target other than an individual Officer or Messenger, they must take it.

If an officer or messenger is in the path of fire, the player who fired (friendly or otherwise) shall roll on Table M to see if there is a possibility of a hit. If both players fire, the enemy player rolls.

This rule was adopted to prevent players from placing their officers anywhere they wish in order to maintain command and control. This rule forces the player to recognize the hazards faced by lone officers and messengers riding across the battlefield.

Compute all fire modifiers as normal, subtracting a -4 for an individual. If the modified die roll is equal to or greater than "5" consult Officer & Messenger Table M and roll again.

An unattached officer or messenger that is not in line of sight is not susceptible to being targeted.

An individual officer or messenger may become the victim of a charge if they are caught between the charging unit and the target of the charge or as a result of pursuit. An officer or messenger can never be the prime target of any charge.

An officer that is caught in a charge is given the opportunity to evade as a skirmisher. If he cannot evade, he is overrun by the opposing unit and taken prisoner.

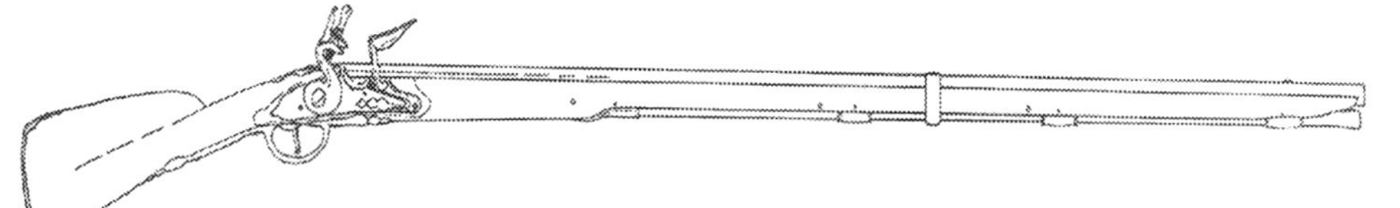

Morale

Units that suffer casualties, declare charges or are being charged may be subject to a morale check.

All D or E class units that take casualties must take a morale check. A, B, or C class units do not have to make a morale check until they reach 25% casualties.

However, once any class unit reaches 25% casualties, it must check its morale every time it receives casualties thereafter. The player checking morale rolls 2d6 factoring in the morale modifiers to the die roll as described in Morale Modifiers Table N.

If the modified die roll is "6" or greater, the unit has successfully passed its morale check and may continue as normal.

Disordered: A modified roll of "5" results in the unit becoming disordered.

Broken: A modified roll of "4" results in the unit becoming broken. A broken unit must withdraw its full unmodified movement directly away from the enemy. At the end of that movement, the units facing is at the discretion of the controlling player.

Routed: A modified roll of "3" or less results in a rout. The unit must make a full rout move directly away from the enemy, ending its movement facing away from the enemy. It must also take one strength point as casualties to reflect desertion, stragglers, etc.

The effects of a morale pass or failure are immediate.

Fear of Disaster

When a unit routs it may potentially affect the morale of the units around it.

A routed unit automatically disorders another unit when it passes through **any** portion of that unit during the first half of its rout movement.

Any unit that was passed through in the first half of rout movement must now take a morale check.

Beaujeu's Last Act

In addition, if any routed unit of any troop-type passes within 6" of an equal or lower class unit during the first half of its rout move, those units must immediately check their morale.

Higher class units are not affected by the rout of lower class units, although they will become disordered if passed through during the first half of rout movement.

Skirmishers rout through other troops without causing disorder. However, if the routing skirmishers are of equal or greater class, the units they pass through must still check their morale for fear of disaster.

Evading skirmishers do not cause a morale check. Only **routed** skirmishers can cause a unit to check morale.

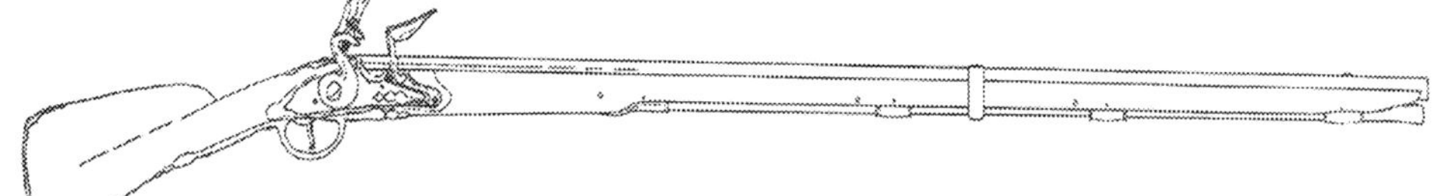

Officer of the Black Watch

Table N. Morale Modifiers

Morale Modifiers	Modifiers
A Class	+2
Stone buildings or fortifications	+2
B Class	+1
Rear support within charge move	+1
Behind friendly lines	+1
Stone fence, palisade, wooden building	+1
C Class	0
D Class	-1
Disordered	-1
25% Casualties	-1
Lost melee this turn	-1
Commander killed or captured	-1
E Class	-2
Broken	-2
Enemy on flank or rear in charge move	-2
Artillery on the flank at normal range or less	-2
Out of Command & Control	-2
Routed	-3
50% Casualties	-4
75% Casualties	-6
Morale rating of attached commander	+ or –

Additional Morale Modifiers for Charges

Opponent Broken or Routed	+2
Charging unit has a 3 to 1 advantage	+2
Charging on Flank or Rear	+1
Charging unit has a 2 to 1 advantage	+1
Being charged	-1
Ambush by previously unseen unit	-1
Europeans charged by Indians for first time	-2
No bayonets vs bayonets or edged weapons	-3
Infantry (non-shock) attempting to charge	-4

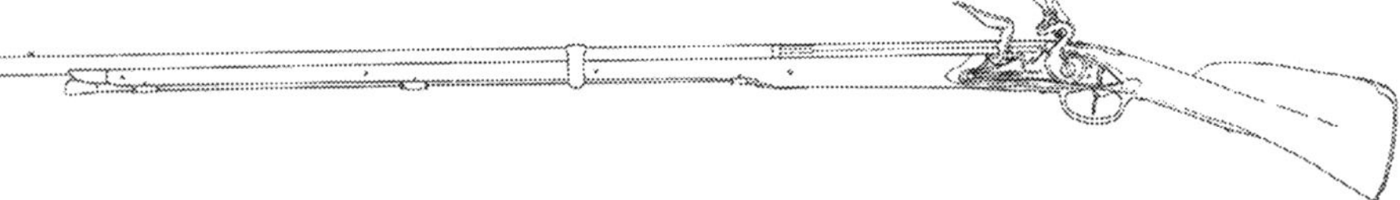

Morale Modifier Descriptions

Most of the morale modifiers are self-explanatory and require little or no description.

Rear & Flank Support: The importance of rear support and secure flanks cannot be overemphasized.

Units with rear support within a charge move receive a +1 bonus to their morale. Conversely, units that do not have secure flanks receive penalties to their morale.

Units that are behind friendly lines receive a +1 bonus. It is advisable to move units with large amounts of casualties to the rear so that they may not only receive this bonus, but can also then provide rear support for the front-line troops.

Note: Routed troops cannot provide rear or flank support. However, broken and disordered units may.

Additional Charge Modifiers

There are certain modifiers that are used only during the Charge/Countercharge Phase. They are listed in a separate section at the bottom of Table N.

Infantry Charges: Historically, infantry charges were relatively rare in 18th Century warfare. They typically only occurred when there was a distinct advantage. To reflect this, there is a -4 modifier for **non-shock** infantry to attempt to charge.

Therefore, this modifier does not apply to shock troops such as Grenadiers & Highlanders.

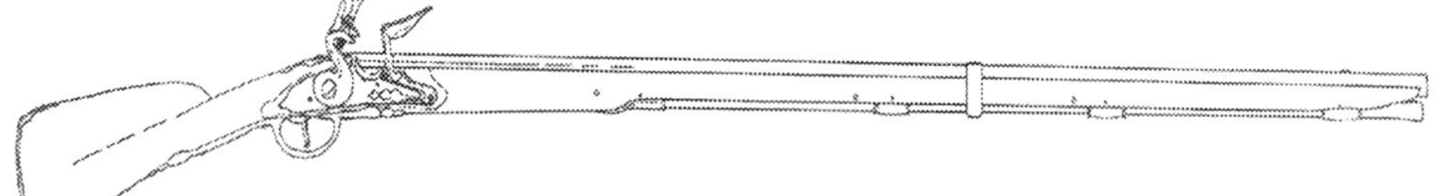

Fate

Fate is the outcome of events regardless of known factors. Fate can be good or bad or have no effect at all.

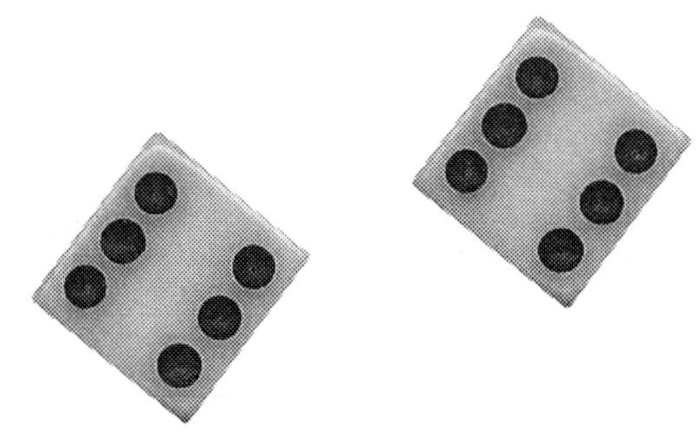

When a player is forced to make a morale check and rolls a natural 2, regardless of the modifiers, the player shall then immediately refer to Fate Table O and roll 2d6 again. Compare the rolled results to Fate Table O.

The Fate Table results are applied to the checking unit.

Effects of the Fate roll are immediate!

Table O. Fate Table

Dice Roll	Results
2	Cowardice in the face of the enemy, unit goes down one grade level. Routs. Unit takes two casualties. Class E unit leaves the field.
3 & 4	Unit Routs. Stragglers desert. Unit takes two casualties.
5 & 6	Unit Broken. Stragglers desert. Unit takes one casualty.
7	No effect. Original die roll & modifiers stand.
8 & 9	Unit Disordered. Follow Procedures.
10 & 11	Unit Rallies. Follow Procedures. Replace one prior casualty.
12	Unit Rallies. Unit recognized for valor. Goes up one grade level. Replace two prior casualties.

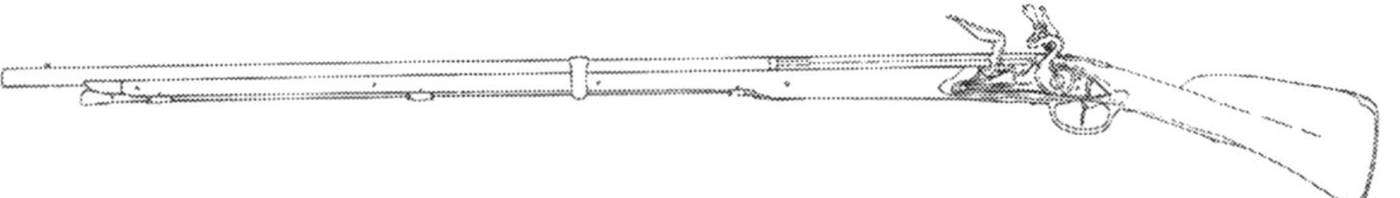

Melee

Contact between two or more opposing units during the charge/countercharge phase results in a melee. Melee combat is resolved in a similar fashion to fire combat.

This chapter is divided into the following sections:

- Melee Dice Determination
- Melee Modifier Descriptions
- Casualty Determination
- Melee Results
    - Melees Resulting in a Tie
- Multiple Units in a Melee
    - Joining an Ongoing Melee

Melee Dice Determination

The number of Melee Dice each unit rolls is determined by that unit's strength factors.

Each stand with four or more figures equals one strength factor. Stands with three figures or less, i.e. Skirmishers, are equal to ½ of a strength factor. Fractions are rounded up. All units roll a minimum of 2d6 and a maximum of 4d6.

Once the number of Melee Dice are determined, the player then rolls those dice factoring in the melee modifiers as listed in Melee Table P.

Melee Modifier Descriptions

Attacked in Rear: Units that attack the rear of another unit do so at a +2 bonus. A unit being attacked in the rear defends itself at -3 penalty.

Table P. Melee Modifiers

Melee Modifiers	Modifier
Vs. Rear of unit	+2
3 to 1 odds	+2
Formed Unit vs. Skirmishers or Open Order	+2
A Class	+2
Charge bonus	+1
Each Strength Factor over 4	+1
Vs. Flank of any unit	+1
2 to 1 odds	+1
Cavalry vs. foot	+1
Highlanders	+1
Indians in the woods	+1
B class	+1
C class	0
Only 1 Strength Factor	-1
Enemy uphill	-1
Vs. Fences, wooden palisades, or buildings	-1
D class	-1
Disordered	-2
Vs. Fortifications or stone buildings	-2
No bayonets vs. bayonets or edged weapons	-2
E class	-2
Broken	-3
Unit attacked in the rear	-3
Attached leader	+ or -

Charge Bonus in Melee: Units that charged that turn receive a +1 bonus.

This bonus only applies during the turn in which the unit charged. This represents the impetus of the charge. It does not apply in a continuing melee.

Highland Units in a Melee receive a +1 bonus. This represents their tendency to drop their muskets and wield their claymores. Thus, they became a formidable fighting force in hand-to-hand combat.

Conversely, once a Highland Regiment engages in melee, they will not have the ability to fire unless they spend their entire movement on the following turn to retrieve their muskets.

Note: If the Gamer chooses not to retrieve his Highlander's muskets, that unit will not be able to fire for the remainder of the battle, or until it returns to the scene of the melee to retrieve them.

Native Americans in a Melee: Indians never willingly entered a melee unless they held a distinct advantage. Therefore, they must hold at least a 2 to 1 advantage over the enemy in order to melee. If they do not, they will choose to evade if possible.

Note: Indians will defend their family and homes as well as defend themselves if cornered. They may also choose to stand and fight if they have cover.

Indians were excellent wilderness fighters. Therefore they receive a +1 melee bonus when fighting in the woods.

If Indians win a melee, there is a 50% chance that they will spend the next full turn scalping and looting. The die roll to determine this will be made immediately after resolution of the melee.

Casualty Determination

The player then divides the modified number by "5" to determine the number of casualties actually inflicted. Every multiple of "5" will result in a casualty. The remainder is discarded.

For example, if a modified roll results in a "12", two casualties would be inflicted on the opposing unit.

Melee Results

The unit that inflicts the most casualties is considered the victor. The loser must withdraw.

The unit that lost the melee must take a morale check **prior** to withdrawing. If he successfully passes his morale, the loser **must** withdraw at least ¼ of its movement away from and facing the enemy.

Disordered: A modified roll of "5" results in the unit becoming disordered. It must move back ½ of its normal movement.

Broken: A modified roll of "4" results in the unit becoming broken. A broken unit must withdraw its full unmodified movement directly away from the enemy. At the end of that movement, their facing is at the discretion of the controlling player.

Routed: A modified roll of "3" or less results in a rout. The unit must make a full rout move directly away from the enemy, ending its movement facing away from the enemy. It must also take one strength point as casualties to reflect desertion, stragglers, etc.

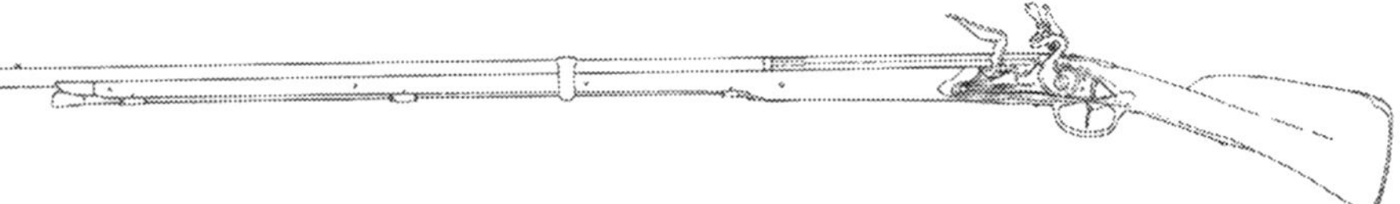

The victor may then occupy the ground vacated by the retreating enemy. All units that fought in a melee are considered disordered at the end of the melee phase.

Melees Resulting in a Tie

If the melee results in a tie, the melee continues to the next turn. All units that had been engaged in the first turn, will be disordered prior to the second turn of melee.

If a melee results in a tie and it involves a unit that is attacked in the rear, that unit can turn around to face the enemy during the second round of melee.

If the second turn of melee results in a tie, **all** engaged units must take a morale check and if successful must withdraw at least ¼ of their movement facing the enemy and are disordered.

If a unit's morale check is unsuccessful, follow morale results procedures.

Multiple Units in a Melee

When a unit that has three or more strength factors is involved in a melee with two or more enemy units, it must split its strength factors between the units and fight a separate melee against each.

If a unit has only two strength factors, it must choose which enemy unit to fight.

If any unit involved in a multiple melee loses one or more of the melees, it must take a morale check.

If the losing unit fails its morale check and is surrounded so that it cannot withdraw it is considered captured. There must be a 3-inch gap between enemy units for the withdrawing unit to escape and avoid capture.

Joining an Ongoing Melee

Units cannot charge into an ongoing melee. They join the melee merely by walking in. They do not receive the charge bonus.

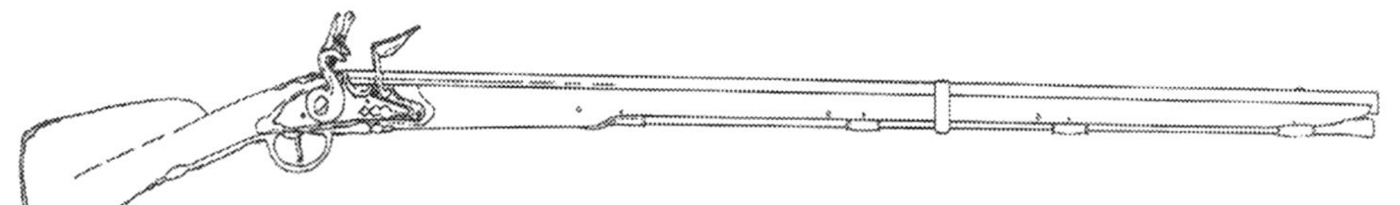

Artillery Reload

Players commanding artillery units shall have the choice of two types of ammunition: canister or ball. Since artillery units would not have traveled in the field with their artillery loaded, they will be considered unloaded when limbered.

When an artillery unit unlimbers, the player commanding that unit shall immediately designate what type of ammunition to load the piece with. This shall be the only exception to the artillery reload phase.

In the turn that an artillery piece fires, the player commanding that unit shall designate what type of ammunition to reload with. This choice shall take place during the artillery reload phase.

It shall be the responsibility of the player commanding an artillery unit to keep track of what type of ammunition has been designated for each artillery piece.

As it will be advantageous for the artillerist to keep this information private until used, the player shall secretly keep track of what ammunition has been loaded into each artillery piece under his command.

This can be done by means of either recording it on paper or by placing a chit upside down next to each artillery piece until that unit is going to fire, then turning the chit over.

The above chits would be designated "B" for ball and "C" for canister.

If a player commanding an artillery unit forgets or fails to designate a particular type of ammunition during the reload phase, the artillery unit shall be considered to have been loaded with ball.

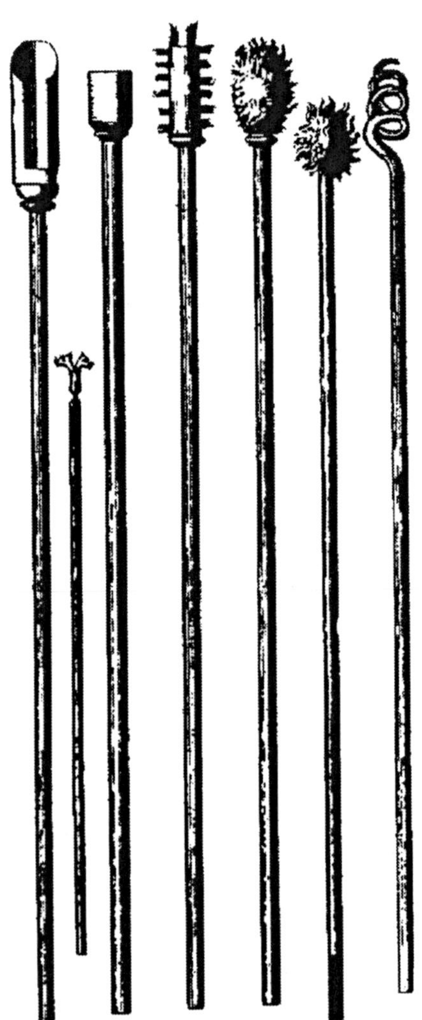

Artillery Tools

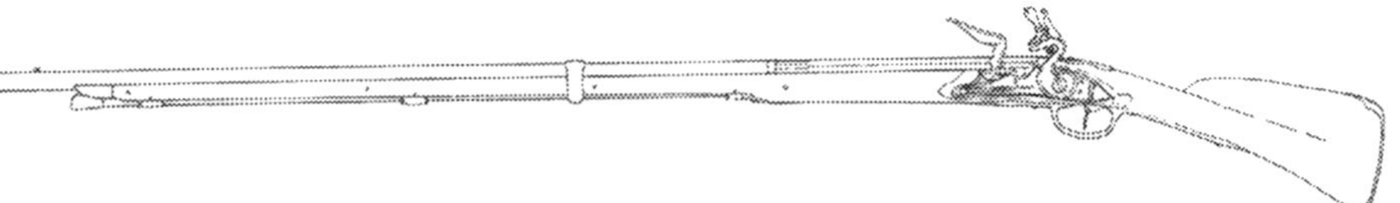

Rally & Recovery

In this phase units that are broken or routed may attempt to rally. Some disordered units may rally at this time as well.

Disordered Units

There are two types of disorder: Disorders caused by movement through terrain and disorders caused by failed morale checks.

Disorders that result from movement through terrain are removed in the Rally & Recovery Phase. These units will automatically rally at this time.

Units that become disordered as a result of a failed morale check must expend half their movement and remove the disorder in the Movement Phase.

Note: Charging units that became disordered as a result of a failed morale check in the Charge Phase, are considered to have expended half their movement.

Therefore, if the player should decide not to move that unit, the disorder would then come off during the Movement Phase.

However, if the unit **did** move during the movement phase, it must remain disordered until the next Movement Phase. It may then expend half its movement to remove the disorder.

French Line Forming

Broken Units

Broken units that did not move, change facing, change formation, fire, or take casualties for one full turn will now rally to disordered status automatically.

Routed Units

Routed units **must** have their unit's Commanding Officer or that officer's superior attached in order to attempt to rally.

In order to rally, the routed unit must successfully pass a morale check. Routed units that fail their morale check **now** rout again. They will immediately flee a full rout move, take another casualty, and face away from the enemy.

Routed units that pass their morale check go to broken status and may now change facing.

Exhaustion & Collapse

A command may become exhausted when it reaches 50% of its full strength factors rounded up.

Once a command reaches 50% casualties, that commander must check to see if his command has become exhausted.

Exhaustion Check

Exhaustion is determined by rolling 2d6 and adding or subtracting the commander's rating modifier, then subtracting -1 for each casualty the command received that turn.

If the modified die roll is equal to or greater than "5", the command passes the exhaustion check. It may proceed as normal. If less than "5", it becomes exhausted.

Exhausted units may not advance towards the enemy unless they are behind friendly lines.

However exhausted units may defend themselves.

Once a command reaches 50% casualties it must check for exhaustion on every turn thereafter that it suffers any casualties.

A command that suffers 75% casualties is automatically considered to be exhausted.

Any unit, in a command that is exhausted, cannot count as rear support.

Note: Artillery unit casualties do not count toward a command's exhaustion nor collapse.

Collapse Check

Once a command has become exhausted it must make a collapse check on every turn that it takes casualties. Again, this is done by taking a morale check with **only** the modifiers as listed in this Chapter above.

When a command reaches 75% casualties it must factor in an additional -2 modifier.

If the command passes the collapse check, it must check for collapse on every turn thereafter that it takes casualties.

All units in a collapsed command go permanently disordered. Any broken or routed units in a collapsed command are removed from the table.

Any units in a collapsed command that subsequently break or rout are also removed from the table.

Army Withdrawal

Once 50% of an army has collapsed it must withdraw from the battlefield. This is designed to prevent the "fighting to the last man" scenario.

The Last Man

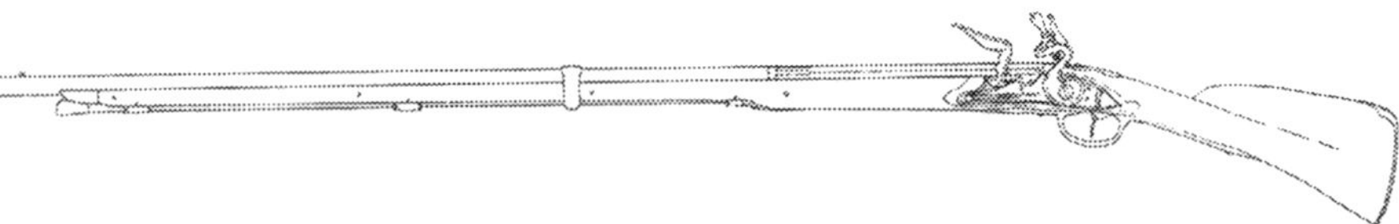

Optional Fortification Rules

Artillery vs Fortifications

The Game Judge can assign strength points to fortifications. Troops within fortifications can only be **directly** fired upon at close range.

In order to inflict casualties on troops within fortifications at normal range or longer, those fortifications must first be destroyed or breached by artillery.

There are many different types of fortifications such as stone walls, earthworks, and wooden palisades.

Stone walls, houses, and buildings obviously represent the strongest type of fortification. Each three inch section would typically be assigned 10 to 20 strength points.

Earthworks, above ground reinforced trenches, wooden forts protected by glacis, and redoubts would be assigned 5 to 15 strength points.

Wooden buildings and wooden palisades would be assigned 3 to 5 strength points.

Once the strength points for a particular 3 inch section have been eliminated, that section is considered destroyed.

Units within the fortification may then be fired upon by both small arms and artillery at all ranges.

In addition, walls considered to have been breached are now subject to direct assault by charging infantry.

Troops behind walls or in trenches can be fired upon by indirect fire from Howitzers or Mortars.

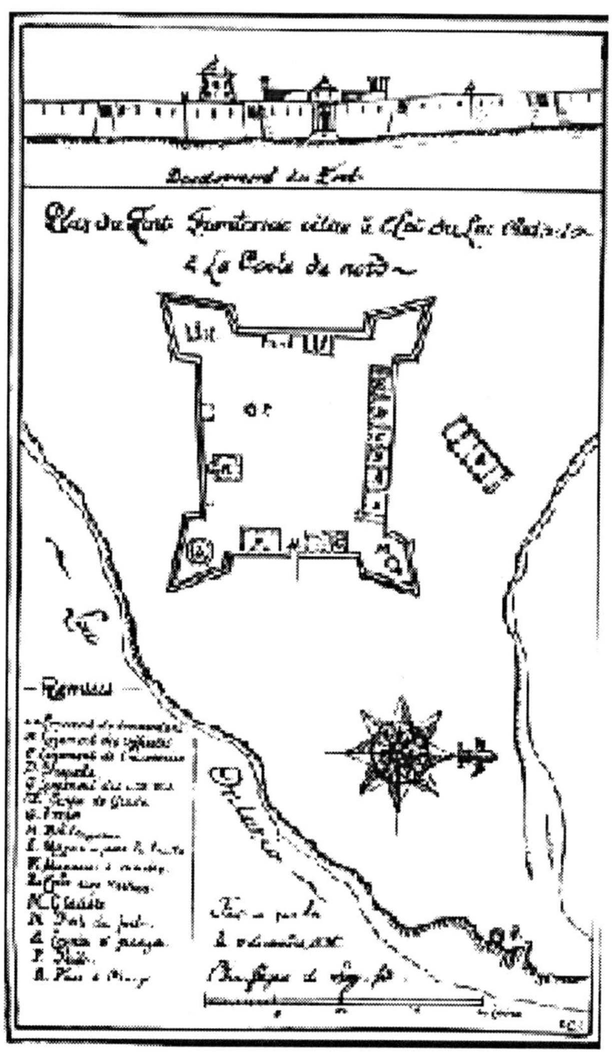

Plan of Fort Frontenac

Turn Sequence

1. Initiative
2. Charges/Countercharges
3. Normal Movement
4. Command & Control
5. Fire
6. Morale
7. Melee
8. Melee Results
9. Artillery Reload
10. Rally & Recovery
11. Army Exhaustion.

Base Movement Rate Table

Formation	Infantry	Cavalry
Line	12"	24"
Column	16"	30"
Line Charge	16"	36"
Column Charge	20"	42"
Skirmish	24"	36"
Road Column	24"	54"
Rout	12" + 2 die	24" + 4 die
Skirmish Rout	12" + 4 die	24" + 4 die

Terrain Effects

Terrain type	Irregular Skirmish	Regular Skirmish	Regular Infantry	Cavalry	Artillery
Light Woods	NE	½	½	¾	¾
Heavy Woods	½	¼	¼	N/A	N/A
Low Rolling Hills	NE	NE	NE	NE	NE
Steep Hills	¾	½	½	½	½ up ¼ down
Shallow Water (below the knee)	¾	¾	¾	¾	¾
Fordable Water (above the knee)	½	½	½	½	½
Deep Water			Impassable		
Marshes	½	½	½	N/A	N/A
Bridges	NE	½	½	½	NE
Towns	NE	½	½	½	NE
Linear Obstacle	¾	¾	¾	½	1 turn
Rough Ground	¾	¾	¾	½	½
Disordered Unit	½	½	½	½	½
Broken Unit	¼	¼	¼	¼	¼

Artillery Movement Rate

Size	Field	Road	Prolong
3 pounds or less (light)	20"	28"	5"
4 to 6 (light)	20"	28"	4"
7 to 10 pounds (medium)	18"	26"	3"
11 to 15 pounds (heavy)	16"	24"	2"
17 pounds or greater (siege)	8"	14"	Pivot Only

For howitzers and mortars, barrel size is equivalent to poundage.

Fate Table Die Results

2	Cowardice in the face of the enemy, unit goes down one grade level. Unit Routs. Unit takes two casualties. Class E unit leaves the field.
3 & 4	Unit Routs. Stragglers desert. Unit takes two casualties.
5 & 6	Unit Broken. Stragglers desert. Unit takes one casualty.
7	No effect. Original die roll & modifiers stand.
8 & 9	Unit Disordered. Follow Procedures.
10 & 11	Unit Rallies. Follow Procedures. Replace one prior casualty.
12	Unit Rallies. Unit recognized for valor. Goes up one grade level. Replace two prior casualties.

Morale Modifiers

A Class	+2
Stone buildings or fortifications	+2
B Class	+1
Rear support within charge move	+1
Behind friendly lines	+1
Stone fence, palisade, wooden building	+1
C Class	0
D Class	-1
Disordered	-1
25% Casualties	-1
Lost melee this turn	-1
Commander killed or captured	-1
E Class	-2
Broken	-2
Enemy on flank or rear in charge move	-2
Artillery on the flank at normal range or less	-2
Out of Command & Control	-2
Routed	-3
50% Casualties	-4
75% Casualties	-6
Morale rating of attached commander	+ or –

Additional Morale Modifiers for Charges

Opponent Broken or Routed	+2
Charging unit has a 3 to 1 advantage	+2
Charging on Flank or Rear	+1
Charging unit has a 2 to 1 advantage	+1
Being charged	-1
Ambush by previously unseen unit	-1
Europeans charged by Indians for first time	-2
No bayonets vs bayonets or edged weapons	-3
Infantry (non-shock) attempting to charge	-4

Troop Quality Movement Penalties

Troop Class	A	B	C	D	E
Retrograde	½	½	½	½	½
Oblique	NE	NE	½	½*	½*
Formation Change	¼	¼	½	Full	Full

* Unit becomes disordered   NE = No Effect

Small Arms Fire Ranges

Weapon	Short	Normal	Long
Musket	3"	6"	12"
Rifle	5"	10"	18"

Artillery Fire Ranges

Gun Type	Canister	Normal	Long	Extreme
Swivel, Wall Guns	3"	5"	10"	20"
2 to 4 pounders	10"	20"	40"	80"
5 to 7 pounders	11"	22"	44"	87"
8 to 15 pounders	13"	27"	54"	107"
16 pounders plus	15"	29"	58"	116"
Howitzers	8"	15"	30"	60"
Mortars	N/A	15"	30"	60"

For howitzers and mortars, barrel size is equivalent to poundage.

Small Arms Fire Modifiers

Short Range	+2
British Regular first fire	+2
Vs. Column or Mob	+2
A class Troops	+2
Target is in the open	+1
Vs. Formed troops in line	+1
Each Strength Factor over 4	+1
Vs. Flank	+1
Non-British Regular first fire	+1
Vs. Cavalry	+1
B class Troops	+1
Normal Range	0
Vs. Light woods or wooden fence	0
C class Troops	0
Formed unit moved under ½	-1
Vs. Heavy woods	-1
Highlanders firing	-1
D class Troops	-1
Disordered	-2
Only 1 Strength Factor	-2
Skirmish unit moved more than ½	-2
Vs. trenches & stone fences	-2
Vs. wooden palisades & buildings	-2
Opportunity fire without movement	-2
E class Troops	-2
Broken	-3
Long Range	-3
Formed unit moved more than ½	-3
Vs. Stone buildings or fortifications	-3
Officer or messenger	-4
Opportunity fire with movement	-4

Artillery Fire Modifiers

Siege Gun	+3
Short Range	+2
Canister	+2
Vs. Column or Mob	+2
A class Gunners	+2
Heavy Gun	+2
Target is in the open	+1
Vs. Formed troops in line	+1
Regular first fire	+1
Vs. Flank	+1
Vs. Wooden fence, due to splintering	+1
Vs. Cavalry	+1
B class Gunners	+1
Medium Gun	+1
Normal Range	0
Vs. Light woods	0
C class Gunners	0
Light Gun	0
Firing unit moved	-1
Vs. Heavy woods	-1
Vs. Wooden palisades & buildings	-1
D class Gunners	-1
Swivel & Wall Guns	-1
Disordered unit firing	-2
Opportunity fire without movement	-2
Vs. Trenches & stone fences	-2
Indirect fire (Howitzers & Mortars)	-2
E class Gunners	-2
Broken unit is firing	-3
Long Range	-3
Vs. Stone buildings or fortifications	-3
Opportunity fire with movement	-4
Officer or messenger (canister only)	-4
Extreme Range	-5

Melee Modifiers

Vs. Rear of unit	+2
3 to 1 odds	+2
Formed Unit vs. Skirmishers or Open Order	+2
A Class	+2
Charge bonus	+1
Each Strength Factor over 4	+1
Vs. Flank of any unit	+1
2 to 1 odds	+1
Cavalry vs. foot	+1
Highlanders	+1
Indians in the woods	+1
B class	+1
C class	0
Only 1 Strength Factor	-1
Enemy uphill	-1
Vs. Fences, wooden palisades, or buildings	-1
D class	-1
Disordered	-2
Vs. Fortifications or stone buildings	-2
No bayonets vs. bayonets or edged weapons	-2
E class	-2
Broken	-3
Unit attacked in the rear	-3
Attached leader	+ or -

Wilderness Wars (Normal Warfare)

Officer & Messenger Casualties

Die Roll	Results
1	Hat shot off. No effect.
2	Slight wound. Out for one turn.
3	Serious wound. Out for two turns.
4	Horse shot, falls on rider. Out for three turns. Captured if in melee.
5	Mortal wound. Officer dies after this turn
6	Shot dead. Hacked to pieces if in melee, body unrecognizable.

Artillery Size

Size	Poundage
Swivel, Wall guns	up to 1
Light	2 to 5
Medium	6 to 10
Heavy	11 to 18
Siege or position guns	19 +

For howitzers and mortars, barrel size is equivalent to poundage.

The Combatants

of the French & Indian War

French & Canadians

French Colonial Militia

The French Canadians have been called the "unsung heroes" of the war in North America.[1] They were outnumbered twenty to one by the British colonists, yet nearly managed to preserve their independence. Every Parish throughout the French colonies was expected to raise every able-bodied man from fifteen to sixty years of age to form the militia.

They were typically gathered into companies of up to 200 men, depending on the size of the local male population, and were often allowed to disband in the fall, winter, and spring for planting and harvest.

In Canada, these companies could be organized into three brigades representing the regions around Quebec, Montreal, and Trois Rivières. In reality, they usually operated in independent companies of 50 to 200 men, under the command of officers from the *Troupes de la marine.*[1]

The French colonial militia were competent skirmish fighters, preferring to fight in open order under cover of the forests. They were often banded together with regular soldiers form the *Troupes de la marine* into ad hoc mixed companies. Morale and organization fell drastically when forced to fight in the open or in formed order. Besides Canada, militia from Acadia, Île Royale (Cape Breton Island), the Illinois Country, and Louisiana also fought in the French and Indian War.

French colonial militia are typically mounted four to a stand in companies of one to five stands. While they sometimes fought as formed troops, such as when defending fortifications or alongside that regulars at the Battle of Quebec, they usually fought in skirmish formation.

While in skirmish the stands in the unit should be separated from ½" to 1". French colonial militia have morale grades of C (usually when in mixed companies with *Troupes de la marine*), D, or E.

Compagnies franche de la marine

Administration of the French colonies fell under the jurisdiction of the Ministry of Marine. Hence the garrison troops of the colonies were recruited, trained, and paid under the auspices of the French Navy.

These *Troupes de la marine* were the regular soldiers of the colonies.

They were primarily responsible for defense of the colony, maintaining internal order, and providing garrisons for the numerous forts and trading posts scattered throughout the North American wilderness.

While most of the enlisted men were recruited in France, by the time of the French and Indian War the officer class consisted almost entirely of native-born Canadians.[1]

Troupes de la marine Behind Earthworks

At the start of the war the *Troupes de la marine* were organized into thirty *compagnies franches,* or independent companies of fifty men. In March, 1756 the size of the companies was increased to 65 men each and in March, 1757, the establishment was increased to forty companies.

Also in 1757, Montcalm formed eight companies into a 500-man battalion. A second battalion was subsequently formed in 1760.[1]

In addition to the companies stationed in New France, the colony of Île Royale had a garrison of 24 companies of fifty men each.[1] Most of these were stationed at the fortress of Louisbourg. The colony of Louisiana had an establishment of 36 companies of 49 men each that were spread from Mobile to the Illinois Country. Some of the latter companies were garrisoned in the Ohio Valley during the war.[2]

The *Troupes de la marine* were experienced woodland fighters who could fight exceptionally well in open order alongside their Native American allies and were often combined with the militia to form ad hoc mixed raiding companies. In addition, they were professional soldiers, drilled in eighteenth-century field tactics and could hold their own alongside the regular French troops.

Troupes de la marine are typically mounted five figures to a stand. They can fight in regular or skirmish formation. When in skirmish formation the stands should be separated from ½" to 2".

Troupes de la marine are considered regular troops and therefore receive a first fire bonus. They are normally morale grade C.

Flag of the Troupes de la marine

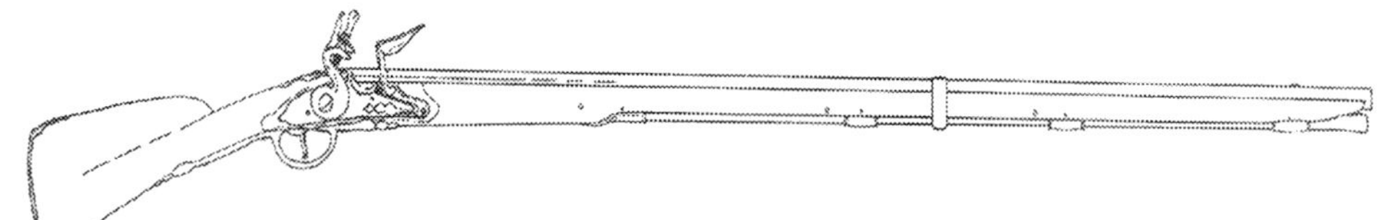

Troupes de terre

The *Troupes de terre* were the regular soldiers of the French Army. In 1755 four battalions of infantry were sent to Quebec and two others landed at Louisbourg. Two additional battalions arrived in Canada in 1756 and two more in 1757. Finally, two additional battalions were sent to Louisbourg in 1758.

Each battalion of French infantry consisted of twelve companies of fusiliers, one company of grenadiers. These companies usually numbered 40 men each.[1] The grenadiers were the elite of the battalion—the biggest, strongest, and most experienced soldiers.

Small platoons of soldiers were often detached for picquets and were used as scouts and could fight in skirmish order. The grenadiers of multiple regiments were sometimes combined to form an elite battalion, although not as often as in the British Army.

Isolated in the New World and blockaded by the British fleet, it became difficult to find qualified reinforcements to fill the void left by casualties and desertion. Consequently, as the war progressed many local militia were recruited into the *Troupes de terre.* This had the effect of significantly lowering the combat capabilities of these units. By the end of the war these recruits represented almost fifty percent of the fighting men of each battalion.[1]

The *Troupes de terre* are considered regulars and are mounted five to a stand. At full strength, each battalion would contain twelve stands of fusiliers and one stand of grenadiers. However, one of the twelve stands could be substituted by three stands of light picquets mounted two to a stand, who would act as a light company.

Troupes de terre were rarely at full strength, and would more often than not only consist of six to eight stands of fusiliers. As regulars, all would receive a first fire bonus. The fusiliers and picquets would typically be morale grade C. The grenadier companies would be morale grade B. They would charge and melee as shock troops.

Troupes de terre Defending

Artillery

The French had two forms of artillery in New France. First there was the colonial artillery of the *cannoniers-bombardiers.* They were the artillery attached to the *compagnies franches* and participated in nearly every action in which the French used artillery during the war.[1] In addition, small detachments of the *Royal-Artillerie* Regiment of the French Army were sent to the colonies along with the regular troops.[2]

French artillery are mounted four artillerists to a gun model and two artillerists for each coehorn mortar and swivel gun. The *cannoniers-bombardiers* manned the vast majority of French artillery. They were considered the elite company of the *troupes de la marine* and as such would have a morale grade B. The *Royal Artillerie* would have a morale grade of C. Both would receive a first fire bonus.

Howitzer

Coureurs de bois

The *coureurs de bois,* "runners of the woods", or bushlopers, were the middlemen of the French fur trade. They lived alongside Native Americans with whom they traded and adopted their way of life. In fact, many were "half breeds", the children of unions between French traders and Native women. They were influential in recruiting the Native American tribes, serving as interpreters and scouts, as well as sometimes even leading them into battle. They typically fought in open order along with their Native American allies.

Coureurs de bois are mounted two to a stand. A unit could number anywhere from one stand to six stands. They would fight as skirmishers and could have a morale grade of B, C, D, or E.

Corps de cavallerie

The *Corps de cavallerie* was the only mounted unit raised in New France during the war. It was formed in June, 1759 of Canadian enlisted men and officers of the regular army. It was used primarily to deliver dispatches but was involved in some skirmishes with British outposts. The *Corps de cavallerie* numbered approximately 200 men.[1]

The *Corps de cavallerie* consists of ten stands of 2 light horse figures each.

British & Colonials

Provincial Regiments

Most of the British colonies raised and maintained their own provincial regiments. Many of the provincials were raised in the cities and served for a contractual period, typically a campaign season.[1] One exception to this rule was the Virginia Regiment, in which most of the troops enlisted for an indefinite period of time.

The provincial units were notorious for their poor discipline and high rate of desertion. Most of the field officers received their commissions from the colonial governors. Morale and level of training varied greatly between regiments. Typically, those regiments under a more permanent establishment, such as George Washington's Virginia Regiment, were more disciplined and better trained than the regiments that were raised on a temporary basis.

Provincial Regiments were usually organized in companies of 50 to 100 men each and could number anywhere from 100 to 1,000 men. They are mounted five figures to a stand. Provincial units can be morale grade C, D, or E. One advantage provincial units have over British regular units is that they may move through rough terrain or woods without going into disorder.

Braddock as He Lay Dying in a Wagon

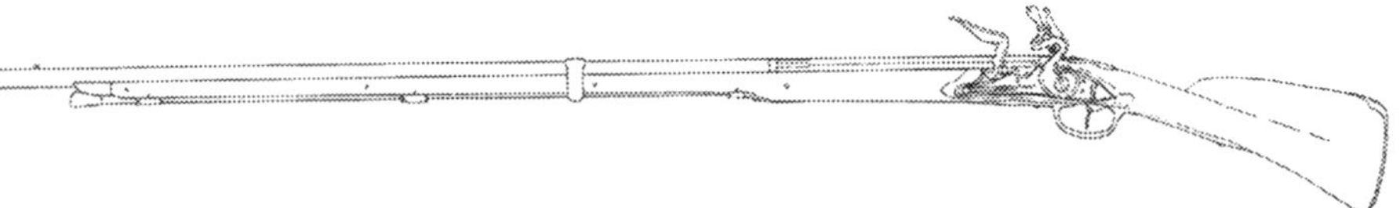

Militia

In addition to the provincial regiments, each colony organized and maintained some form of local militia. These units usually consisted of the local adult males of a county or militia under the command of local leading citizens.

Their formal training usually consisted of mustering on the local common once or twice a year. Often, these musters quickly devolved into a social gathering where little or no military training actually occurred.

The militia were sometimes called up on a relatively limited basis in times of local emergency, such as during period of heavy Indian raids on the frontier. Once the immediate threat had passed, they were usually dismissed. They were often of little or no military value. As Lieutenant Colonel Adam Stephen of the 1st Virginia Regiment once wrote, "the Militia will add to our disgrace but nothing to our Strength."

The militia of the British colonies would be considered very poor quality with a morale grade of D or E. They are typically mounted 4 men to a stand. Militia units could number anywhere from 40 men to battalions of 500 men. They can remain formed or go into skirmish.

Sometimes, during times of attack on the frontier, local settlers would gather in to ad hoc groups to defend themselves. These can be mounted on 4 figure stands and act as militia or they can be mounted 2 figures to a stand. Two-figure stands are always considered to be in skirmish.

British Regiments of Foot

Grenadier

During the course of the war approximately nineteen regular British Regiments of Foot served in North America. Most of these regiments consisted of one battalion with eight companies of musketeers. There were a few exceptions to this rule. One of those exceptions was the 60th Regiment, or Royal Americans, which consisted of four battalions. However, even in this case, the battalions usually acted independently and no more than two battalions ever served together in the same campaign.

Each regiment also contained one grenadier company and one light company that were often referred to as "flank companies." These companies were often combined to form flank battalions of grenadiers or light infantry.[1]

The British Regiments of Foot are considered regular units and are mounted 5 to a stand. At full strength, each battalion would contain eight stands of musketeers and one stand of grenadiers. In addition, they would contain a light company of three stands mounted two to a stand.

However, they were rarely at full strength, and would more often than not only consist of six to eight stands of fusiliers. As British regulars, all would receive a +2 first fire bonus. The musketeers and light company would typically be morale grade C. The grenadier companies would be morale grade B and would charge and melee as shock troops.

Independent Companies

The Independent Companies were companies that were specifically formed to serve in garrisons in North America. They were in the pay of the King and were considered regular soldiers. The Independent Companies would usually number 100 men under the command of a captain when at full strength.

They played an important role during the early years of the war and were sometimes formed together into battalions. However, as the war progressed and casualties mounted they were increasingly incorporated into the regular battalions. The last of the Independent Companies in South Carolina and Georgia were disbanded in 1763.[1]

Independent Companies are considered British regulars and thus receive a +2 first fire bonus. When at full strength they are made up of two stands with five figures each. They receive a morale grade of C.

Light Infantry

As the war progressed the British came to realize the importance of light infantry. As such, a few light infantry regiments were formed. The most famous was the 80th Light Infantry, or Gage's Light Infantry. Some provincial regiments were also established as light infantry, such as Partridge's Light Infantry battalion, which served in the Ticonderoga campaign of 1758.[1]

Light infantry regiments are mounted on two man stands and always fight in skirmish order. These regiments could number anywhere from 300 to 900 men. The morale grade would usually be C, D, or E, depending on the quality.

The British regular light infantry would receive a +2 first fire bonus. Provincial light infantry units would not receive a first fire bonus. For purposes of game play, these regiments can be split into smaller groups of five to ten stands, or the equivalent 100 to 200 men.

Highland Regiments

Beginning in 1758, the British began to raise regiments from the Scottish clans. The Scots were intimidating warriors and the Highland Regiments proved to be very effective. They tended to be very large and could number as many as 1,400 men.

They were armed with large Scottish broadswords known as Claymores, which made them extremely effective in melee.

However, they often dropped their muskets when pulling out their swords. Thus, the subsequent firepower of the regiment as a whole would be seriously weakened, such as when the 78th Regiment charged the fleeing French at the Battle of Quebec.[1]

Highland Regiments are mounted on five-man stands. A regiment could number anywhere from 500 to 1400 men. Highland Regiments can move through rough terrain without going into disorder. The Scots were ferocious in hand to hand combat and thus receive +1 melee bonus.

However, due to their tendency to drop their muskets, once they enter melee, their firepower is halved for the remainder of the game. In other words, if the regiment normally has ten strength points, only five points can fire.

Highland Regiments receive the +2 British first fire bonus. Highlanders are normally morale grade B. However, Highland Grenadiers could be morale grade A. All Highland units charge and melee as shock troops.

Artillery

The British sent over many companies of the Royal Regiment of Artillery. In addition, some colonies also had their own provincial artillery companies.

British artillery are mounted four artillerists to a gun model and two artillerists for each coehorn mortar and swivel gun model. The Royal Artillery would have a morale grade of C while provincial artillery would be C, D, or E depending on their quality. Only the Royal Artillery receives a first fire bonus.

Officer & Sergeant

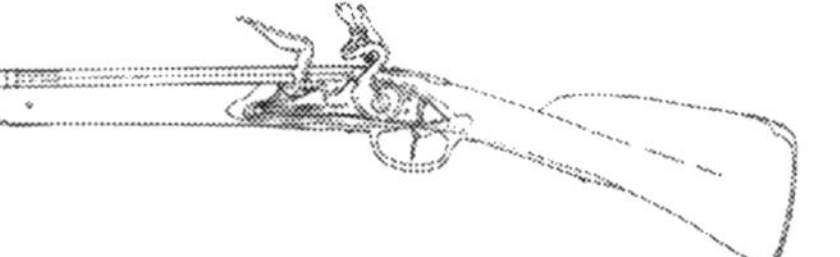

One of Roger's Rangers

Ranging Companies

The British also raised independent ranging companies to serve as scouts. The most famous of these were Rogers' Rangers. Many of the companies, such as Rogers' were placed on the King's payroll. However, the various colonies also raised ranger units. Most of the ranging companies numbered 100 men when at full strength.

Ranger companies consist of five stands with two figures each. They can be of any morale grade depending on their quality.

Cavalry

Cavalry played a relatively minor role in the French and Indian War. A few of the colonies raised small troops of light cavalry at different times during the war but they saw little or no combat. The most visible was the Virginia Light Horse, which served as guard to General Edward Braddock and fought at the Battle of the Monongahela where it was nearly wiped out.

After 1755, the Virginia Light Horse was relegated to garrison duty and was eventually merged into the rest of the Virginia Regiment.[1] No Regular British Cavalry Regiments served in the French and Indian War.

Provincial cavalry regiments would consist of one to three stands with two figures each. Their morale grade would be C, D, or E, depending on their quality.

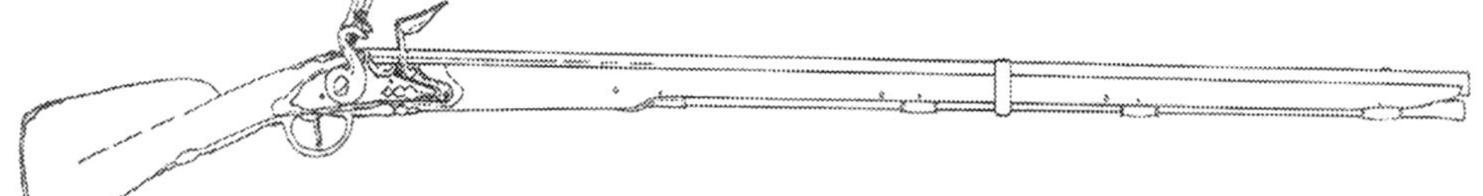

Native Americans

The Native American warriors were excellent woodland fighters. Often brutal and cruel, not only did they fight as auxiliaries with the opposing armies, they also participated in war parties, raiding the frontier settlements and the villages of natives allied with the opposing side.

Most of the Native Americans in the eastern portion of North America eventually allied themselves with the French, seeing them as least objectionable of the European powers.

The tribes can be divided into several groups—Those tribes traditionally allied with the French included the Algonkins of Canada, the Abenaki of New England, and the Micmacs of Acadia. These nations had long been the enemies of the English colonials and had been conducting raids against the New England and Acadian frontier settlements throughout most of the Eighteenth Century.

The Hurons, located on the northeastern shores of the Great Lakes were also closely linked with the French and had long participated in their wars against the Iroquois and the British.

The Western Indians of the Great Lakes region and the Illinois also had close ties to the French but had not been involved in their wars with the British until recently. They actively fought with the French during the early years of the war.

The Indians of the Ohio Country had a long history of trading with both the British and French. Most of the Delaware, Shawnee, and Mingo tried to maintain their neutrality or openly supported the British up until the Battle of the Monongahela in July, 1755.

A Miami of the Ohio Country

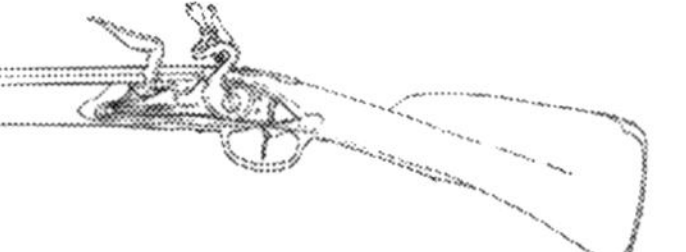

After Braddock's defeat, most of these tribes went to war against the British, fearing reprisals from the French if they attempted to maintain their neutrality. These tribes were among the first to make peace once the British began to gain the upper hand.

The Iroquois Confederacy, the most powerful Native alliance in North America attempted to officially maintain neutrality throughout the war.

However, the Seneca and many in the other tribes of the Confederacy actively supported the French while the Mohawk, the tribe closest to the British settlements in New York, actively supported the British. There was also a large number of Christianized Iroquois, commonly known as the Caughnawaga, who had moved to Canada. They were staunch allies of the French.

Many of the Southern tribes fought with the British. The largest and most influential of these were the Cherokee. Cherokee warriors participated in many raids against the French and their Indian allies in the Ohio Valley and on the frontiers of Virginia, Maryland, and Pennsylvania. The Cherokee soon became disillusioned with their treatment by the British. This resulted in open conflict during the Cherokee War of 1760-61.[1]

The Native Americans almost always fought in loose bands of from 50 to 200 warriors, usually under the leadership of a prominent war chief. When fighting with the French *coureurs de bois* or officers of the *Troupes de la marine* would often accompany them as interpreters. The British also had Indian officers, who would lead or accompany war bands in the field.

Mohawk of the Iroquois Confederacy

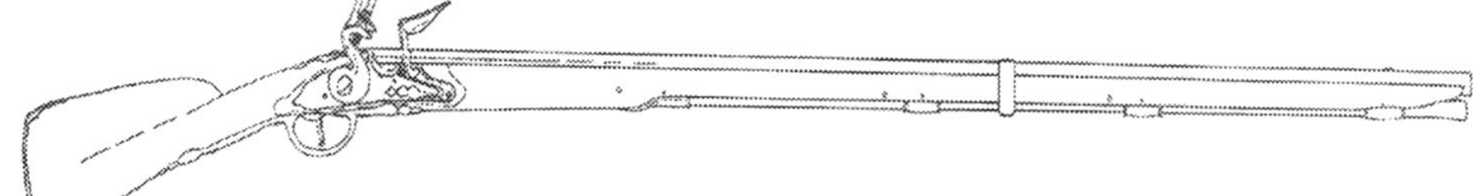

They were excellent woodland fighters and marksmen, who often targeted officers and other prominent enemy figures on the battlefield.

However, their morale diminished drastically when forced to fight in the open or against fixed positions. They were also often unruly, and were prone to stop fighting in the middle of a battle to loot or collect scalps.

Finally, they were costly to maintain as they expected to be clothed, fed, armed, sheltered for their services.

Native American warriors are mounted on stands with two figures each. They receive a +1 bonus to their morale while in the woods.

Native Americans will not voluntarily join in melee unless they outnumber their enemy 2 to 1. They preferred to fire and retire unless they had a clear advantage.

The exceptions to this rule would be that they are surrounded or are defending their village and families. However, Indians that are in hand to hand combat in the woods receive a +1 melee bonus.

If Native Americans win a melee there is a 50% probability that they will stop one turn to collect scalps and booty. Native Americans can be any morale grade depending on their quality.

End Notes

[1] Christopher Duffy, *Warfare in the Age of Reason: 1715–1789,* (New York: Barnes and Noble Books, 1997), p. 282.

[2] René Chartrand, *The French Soldier in Colonial America.* Historical Arms Series, No. 18 (Bloomfield, Ont.: Museum Restoration Service, 1984), p. 9.

[3] Ibid., p. 10.

[4] Ibid., pp. 11-14.

[5] Ibid., pp. 15-16.

[6] Ibid., pp. 16-19.

[7] Ibid., p. 30

[8] Ibid., p. 32.

[9] Ibid., pp. 27-28.

[10] Ibid., p. 30.

[11] Ibid., p. 36.

[12] Fred Anderson, *A People's Army: Massachusetts Soldiers and Society in the Seven Years' War,* (New York: W. W. Norton & Company, Inc., 1985), p. 27.

[13] "Adam Stephen to George Washington, July 25, 1756," in W. W. Abbott, Dorothy Twohig, et. al., eds. *The Papers of George Washington: Colonial Series.* 10 Vols. (Charlottesville, VA: University Press of Virginia, 1983-95), 3:295-96.

[14] Stuart Reid, *King George's Army 1740-93: (1) Infantry.* Men-at-Arms Series, No. 285, (London, Osprey Publications, 1995), pp. 5-9.

[15] "Disposition of Troops, August, 1763," in Clarence W. Alvord, *The Critical Period 1763–1765.* Collections of the Illinois State Historical Library, Volume X (Springfield, Ill., Illinois State Historical Library, 1915), pp. 14-17.

[16] René Chartrand, *Ticonderoga 1758.* Osprey Campaign No. 76 (Oxford: Osprey Publishing Ltd., 2000), p. 27.

[17] René Chartrand, *Quebec 1759.* Osprey Order of Battle Series, No. 3 (Oxford: Osprey Publishing Ltd., 1999), pp. 92-93.

[18] "Robert Dinwiddie to George Washington, August 19, 1756," *The Papers of George Washington,* 3:358–62.

[19] John Richard Alden, *John Stuart and the Southern Colonial Frontier* (New York: Gordian Press, Inc., 1966). Chapters 6–8 provide an overview of the breakdown of the Anglo-Cherokee Alliance and the subsequent outbreak of hostilities.

76
ION
DENCE

The Combatants

of the American War of Independence

By Forrest Harris

The following article outlines ways to organize your American Revolution armies to fight battles using *Wilderness Wars*.

Style of Fighting

The historical record does not support the popular myth that Americans won the war with a handful of wily riflemen shooting red-coated, wooden soldier-boys from behind trees and rocks.

Although partisan tactics were employed in backwoods areas, the war was won only when Americans, with the help of officers like the famous drillmaster Baron Von Steuben, finally achieved a level of expertise and discipline to go toe-to-toe with British regulars in the open.

Veteran regiments from Maryland and Delaware proved their mettle in the latter years of the war by thrashing His Majesty's best regiments with cold steel.

This is not to suggest that the war was conducted in the same manner as in Europe. Much had been learned since the French and Indian War and the British, in particular, had made significant strides in adapting their tactics to American battlefield conditions.

First, and most significantly, the American landscape, without large expanses of groomed fields, was not good heavy cavalry country, and the light cavalry was not suitable for frontal assaults.

As a consequence, dense infantry field formations were not needed, and the King's units adopted what was called the "loose manner" of forming.

This formation was a flexible two-rank line in which the files were staggered in a zigzag pattern, formed by every other man in a single line stepping backward one pace.

The intervals between men could be quite large so that the formation resembled a skirmish line. This should be considered the "default" formation for most troops by the end of the war.

In the American Revolution, armies were organized into Brigades of between two and four regiments (even when only one battalion was present, the unit was called a "regiment" in the parlance of the day).

The Brigades were organized into "lines," and not "divisions" or "corps," at least not as the term was to be used in later wars.

The army was generally deployed with one line under one commander in the front and another line under another commander behind the first.

A "corps" of light infantry at brigade strength was on each flank. Light cavalry was in reserve to exploit enemy weak points or protect the army's flank and rear, depending on how the day progressed.

In the middle of the war, both sides developed "legions," small, elite forces of combined light cavalry and infantry who fought for control of the flanks during a battle. They often appeared on orders of battle as "corps of observation."

Artillery, never in great numbers, was placed most commonly in the center of the line, on a road when possible, or in some advantageous defensive feature such as a steep knoll or redoubt.

Artillery was transported to the battlefield by civilian contractors, dropped off and dragged by the gunners, on foot, with towlines into firing position.

If the position was overrun, the civilian contractors would typically flee and the guns were captured. Because of these complications, it was considered no shame to lose one's guns.

Continental Line in "Loose Manner" or Zigzag Formation

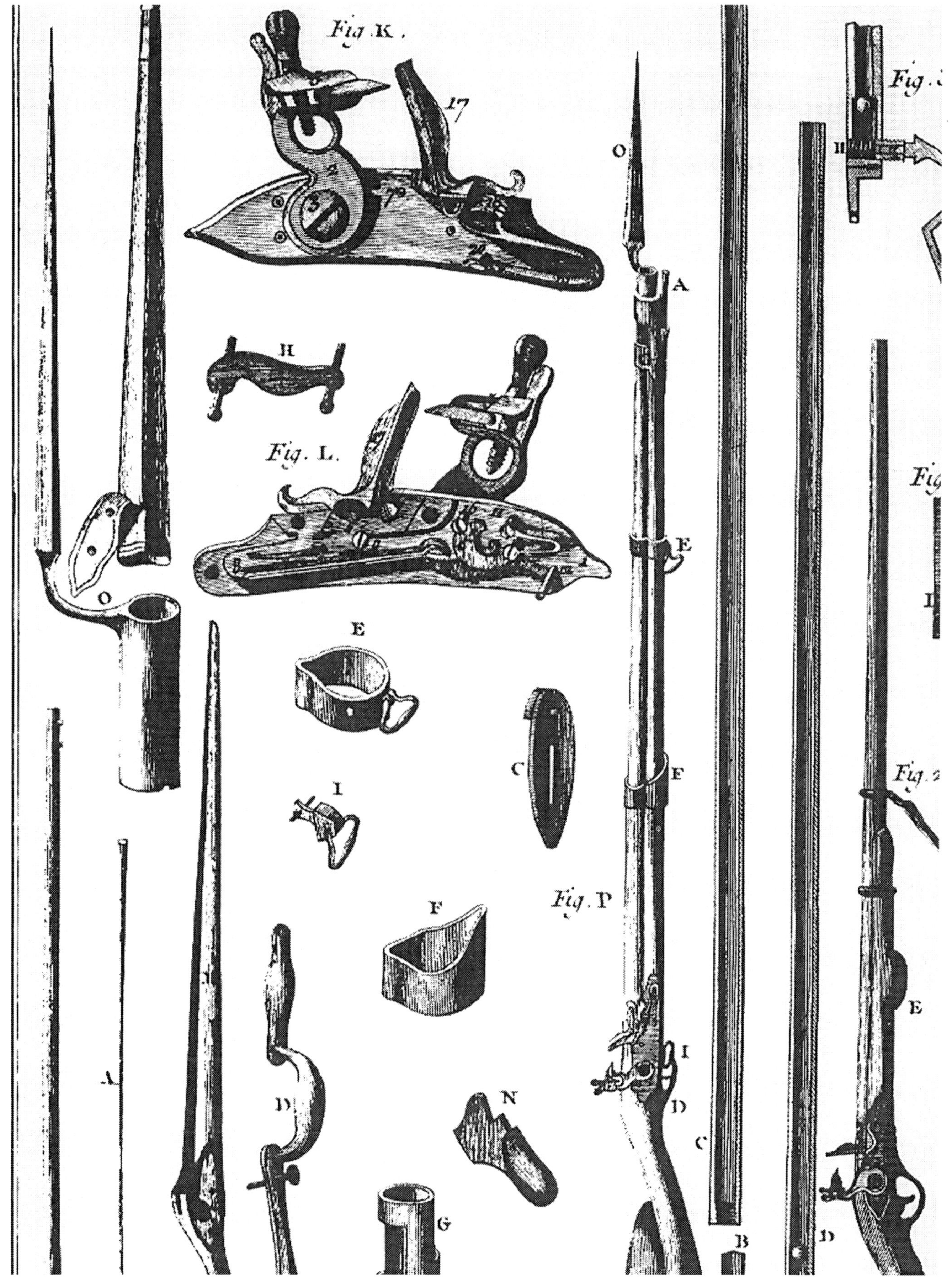

Fig. K.
17
H
Fig. L.
O
E
C
I
F
Fig. P
D
N
G
A
O
A
E
F
I
D
C
B
D
E

Americans

Continental Line Regiments

Organization: 8 stands of 4 figures each, which can be split into two battalions of 4 stands each. One stand of each regiment is light infantry.

Class: Range from C to A, depending on experience level.

Continental Rifles & Light Infantry

Organization: "Corps" of varying sizes, often times attached to Continental Light Dragoon units to form "legions." Infantry elements can be between 2 and 12 stands of 2 figures each.

Class: Many at B class or higher. Some, like Morgan's Rifles, could be considered A class. Early in the war, many were willing, but not able, and could be rated as low as C.

Continental "Light Horse"

Continental Horse (a.k.a. Light Dragoons) carried firearms and engaged in skirmishing with enemy elements (and were famous for raids against enemy supply convoys), however they preferred the sabre in major actions.

They distinguished themselves at Cowpens and elsewhere for their successful charges against the flanks and rear of British regulars, as well as their victories against enemy cavalry using cold steel.

Organization: A full regiment was 360 dragoons, in 6 troops (36 figures on 18 stands, with three stands forming a troop), but the troops often operated in smaller formations, and were sent on details apart from their fellows.

Continental Line Regiment

Midway through the war, troops of cavalry were combined with light infantry companies to form "legions."

Quality: B class, with a few A class troops, such as Lee's Legion (Light Horse Harry Lee).

Militia & State Troops

The upper echelon of militia consisted of men who enlisted for a term of service, as they were accustomed to from colonial times. They received some modicum of drill instruction (although often antiquated) and some military equipment (usually no official uniforms).

Some of the men were recently-discharged veterans of the Continental Line. They were often supplied with old muskets, but rarely had bayonets, or the training to wield them convincingly.

Organization: Varies, but generally along the same lines as the Continental establishment; from 4 to 8 stands of four figures each.

Quality: C class or lower; often D or E class. The two main problems were insubordination and lack of proper equipment, especially bayonets, the latter of which was a critical inadequacy.

These troops were not able to skirmish, because they were not taught modern light infantry drill. The higher quality state troops, however, were trained and equipped like Continentals and were sometimes upgraded into the Continental line.

Backwoods Militia

Backwoods militia units were often not drilled at all. They arrived to battle on their own horse (which was left in the rear) and left when they felt like it. They carried their own hunting rifles and dressed in frontier garb.

Organization: The basic unit of organization was the company, usually no more than 40 men (2 stands of 2). Companies drifted into camp from distant settlements in the days leading up to a battle.

They were assembled into loose formations about the size of a Continental brigade, but without any recognizable command structure or regimental continuity.

They were usually assigned the defensive of a geographic feature, such as a redoubt, hill, or fence row.

Quality: Usually D or lower, however in the right situation, they performed at much higher levels. At King's Mountain, a band of backwoods militia, fighting as a loose mob, surrounded and annihilated a force of trained Tory regulars.

Artillery

Organization: Artillery was formed into "regiments" of 12 companies, each with two guns and 63 men. For game purposes, one or two stands per "wing" or "line" (equivalent of a division) should be assigned.

Sizes ranged from 3- to 8-pounders commonly in use in the field.

Quality: B or C class, depending on experience level of the battery.

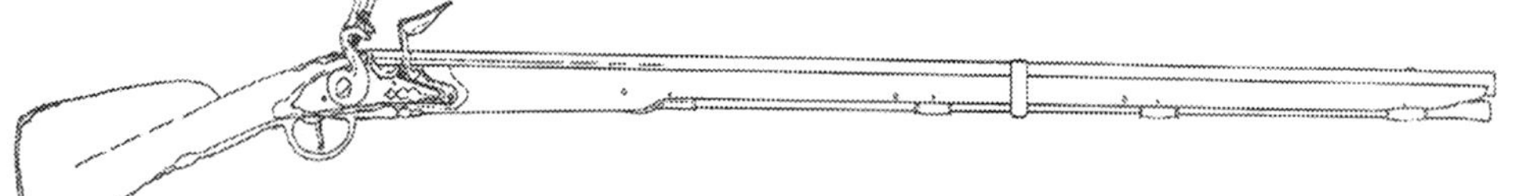

British

Guards

One thousand men were handpicked from among all three Guards regiments in order to form a "Guards Brigade" to send to America.

This Brigade was divided into two Guards battalions (five companies in each, with one battalion containing all of the grenadiers and the other containing all of the light infantry).

Organization: 5 stands of 4 figures each per battalion, which may act together as a unified regiment.

Quality: A class.

Line, Fusilier & Highland Regiments

The British Army of 1776 was perhaps the best man for man of any in history. True, the Fusiliers and Highland troops were elites in theory.

However, here was really not much difference between them and the ordinary line regiments serving in America (except for the Highlanders' peculiarities, which are enumerated in the rules).

Light Infantry, One of the King's 8th

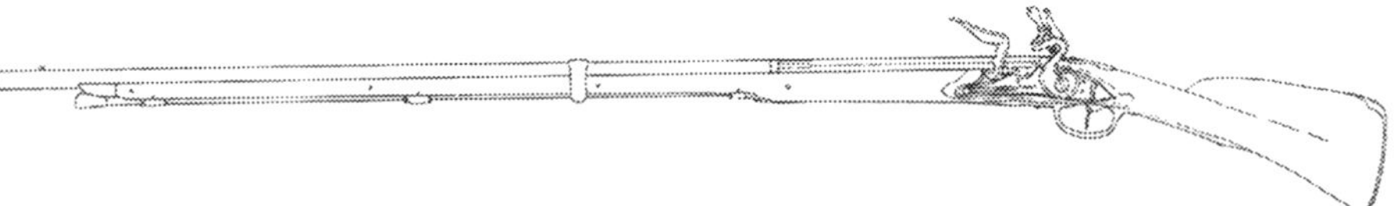

The quality of all the troops sent to America was very high; the Crown could not afford to send anything less than the best for so critical a mission.

Organization: British infantry regiments were organized in ten companies of 60 men each. One of these companies was composed of grenadiers, posted on the right of the line of battle, and one of light infantry, posted on the left. The remainders were "hatmen," rank-and-file regular infantry.

Quality: B Class. The level of training and discipline among regular British line regiments was superb, and their steadiness in battle is legendary.

Note: Orders of battle sometimes mention "light" regiments (as opposed to light companies). This designation can largely be ignored, since all the King's troops were taught the "loose manner of forming" described at the beginning of this chapter.

Tory Militia

Organization: Same as British regulars.

Quality: C to E class, depending on level of experience and training.

Tory Backwoods Militia

Same as American Backwoods Militia.

Light Dragoons

The British employed both Regular and Tory light horse.

Organization: Approximately 240 men in 6 troops, represented by 12 stands of 2 figures each (2 stands to the troop).

Quality: B class for British regulars (16th and 17th Light Dragoons) and C class for Tory cavalry. The vaunted Tarleton's Legion (Tories led by a British commander) was feared for its brutality, but was not always reliable in battle.

Artillery

Organization: Royal artillery was organized into companies of 80 men and 2 guns each, to be represented by one stand on the game table.

One or two companies would be assigned to a "line" or "wing" (division-strength element). All calibers were used, however 3 and 6 pound guns were common in the field.

Quality: B class.

Preparing the Guns

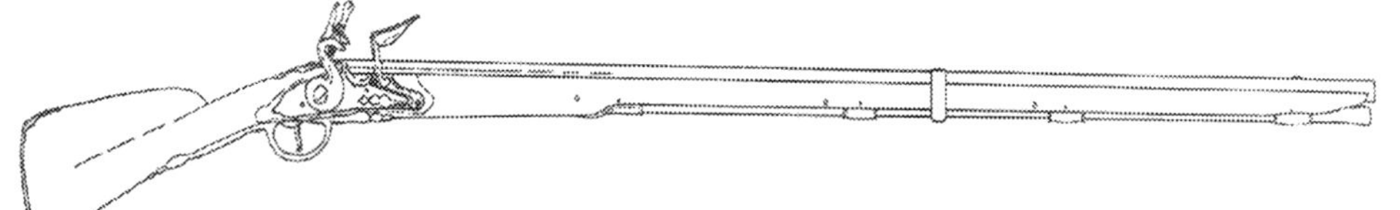

Germans

Hessian Infantry

Organization: 6 companies of approximately 120 men each (5 regulars and 1 grenadier), represented by 18 stands of 4 figures each; 15 stands of regulars and 3 stands of grenadiers.

Quality: B class for grenadiers, C class for regulars. The Hessians never lived up to their reputation.

German Regimental Flag

Jaegers

These rifle-armed German troops operated as skirmishers throughout the war. There were also mounted Jaegers. Jaeger is the German word for hunter.

Organization: Independent companies of approximately 120 men, represented by 6 stands of 2 figures, each.

Quality: B class.

References

The best reference on the subject for Gamers is without a doubt Greg Novak's two-book series The American War of Independence, A Guide to the Armies of the American War of Independence, published by Old Glory.

Other useful references include Uniforms of the American Revolution by John Mollo, and numerous Osprey titles.

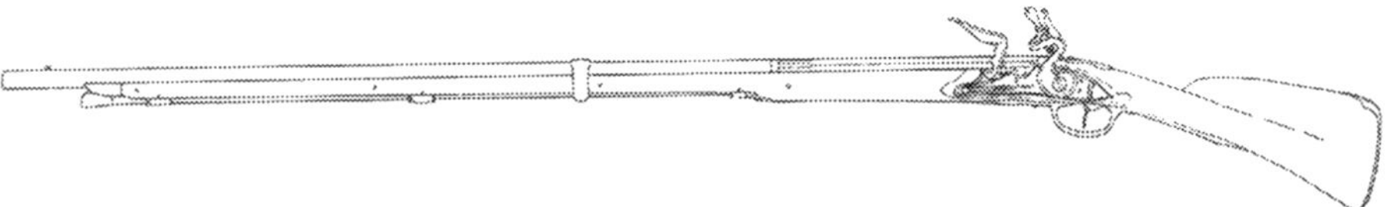

Index

D

E

F

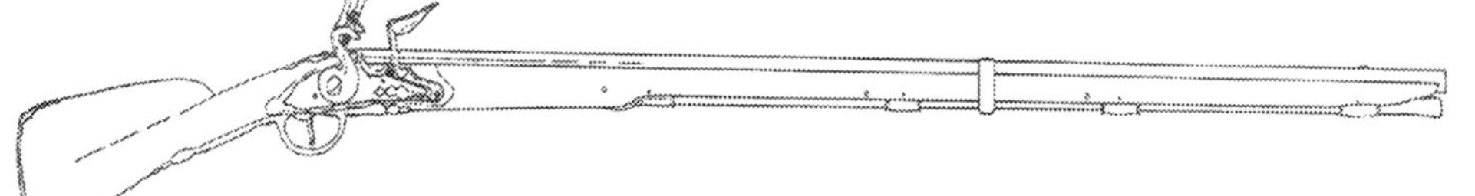

N

O

U

W

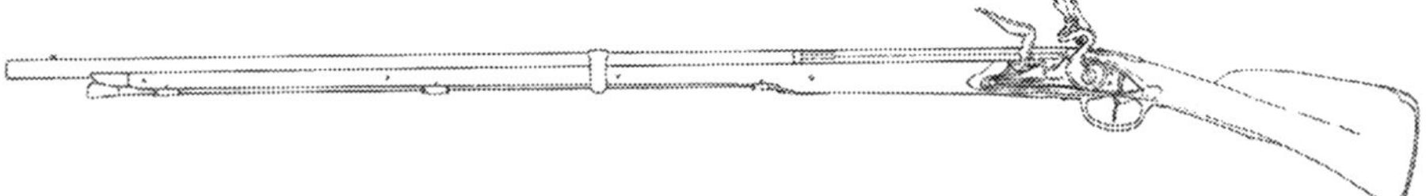

Normal Warfare ©

Come visit us at our website

www.normalwarfare.com

P. O. Box 35, Normal, Illinois 61761

Printed in the United States
62713LVS00006B/14

9 780974 869001